THE SPIRIT'S JOURNEY

DAVE MCKENZIE

IUNIVERSE, INC.
BLOOMINGTON

The Spirit's Journey

Copyright © 2011 by Dave McKenzie

All rights reserved. No part of this book may be used or reproduced by any means, graphic, electronic, or mechanical, including photocopying, recording, taping or by any information storage retrieval system without the written permission of the publisher except in the case of brief quotations embodied in critical articles and reviews.

The views expressed in this work are solely those of the author and do not necessarily reflect the views of the publisher, and the publisher hereby disclaims any responsibility for them.

iUniverse books may be ordered through booksellers or by contacting:

iUniverse
1663 Liberty Drive
Bloomington, IN 47403
www.iuniverse.com
1-800-Authors (1-800-288-4677)

Because of the dynamic nature of the Internet, any Web addresses or links contained in this book may have changed since publication and may no longer be valid.

ISBN: 978-1-4502-7221-6 (sc)
ISBN: 978-1-4502-7220-9 (dj)
ISBN: 978-1-4502-7219-3 (ebk)

Library of Congress Control Number: 2010916498

Printed in the United States of America

iUniverse rev. date: 1/28/2011

in memory of Jewel Currence McKenzie &
George Dewey "Mac" McKenzie

CONTENTS

I.	GEORGE DEWEY MCKENZIE	1
II.	THE BUG BIT	3
III.	JEWELL CURRENCE	14
IV.	DAVID ARRIVES	26
V.	A FINAL MOVE	29
VI.	JAMES LEONARD MCKENZIE	33
VII.	FORMAL EDUCATION FOR DAVID BEGINS	38
VIII.	JIMMY STARTS SCHOOL	54
IX.	DAVID MOVES UP	60
X.	JUNIOR HIGH SCHOOL	70
XI.	LIFE & ENVIRONMENT CHANGES	80
XII.	JIMMY GETS HIS OLD CAR	87
XIII.	CREATIVITY SURFACES	97
XIV.	COLLEGE YEARS BEGIN	109
XV.	SOUTHERN COACH & BEGINNING A CAREER	135
XVI.	ON THE WAY UP AT LAST	142
XVII.	THE MILITARY OBLIGATION	155
XVIII.	LIFE RESUMES	159
XIX.	THE SOUTHERN COACH S-41-DM	166
XX.	CAP, an L-16, and TWO AIRSHOWS	168
XXI.	N-71568 ARRIVES	175

XXII.	A NEW PRODUCTION MANAGER JOINS SOUTHERN COACH	180
XXIII.	IT'S TIME TO MOVE UP	187
XXIV.	THE CLARK CORTEZ LEADS TO A FUTURE	191
XXV.	THE CAREER PATH LENGTHENS	194
XXVI.	STARTING A HOME & LIFE GOES ON	200
XXVII.	THE DESIRE TO FLY IS WELL AND STRONG	205
XXVIII.	A NEW OBJECTIVE	208
XXIX.	BIRDS OF A FEATHER FLOCK TOGETHER	216
XXX.	N-5747-N AND A LONG TRIP TO FLORIDA	220
XXXI.	"CLIPPED WING" CUB, AGAIN	229
XXXII.	A NEW OBJECTIVE (cont'd)	231
XXXIII.	A FEW TALENTED YOUNGSTERS	236
XXXIV.	"THE SIGN", ATC, & A TRIP FROM NASHVILLE	242
XXXV.	THE INTERNATIONAL AEROBATIC CLUB, CHAPTER 88	251
XXXVI.	THE ULTIMATE DREAM	258
XXXVII.	THE PRICE	265
XXXVIII.	CHANGES AND ADVANCEMENTS	268
XXXIX.	THE NEXT THREE YEARS	273
XL.	CAN THESE VEHICLES TRANSCEND TIME?	275
XLI.	LIFE DOESN'T CHANGE	280
XLII.	THEY HAD an AIR SHOW at HOWELL	282
XLIII.	DETROIT CITY AIRPORT SHOW	287
XLIV.	OPEN HOUSE at NEW HUDSON	290
XLV.	FALL CARNIVAL AT CLINTON	292

XLVI.	WINTER – '79 through "80	294
XLVII.	LOST and DISORIENTED	297
XLVIII.	ANOTHER OBSTACLE	307
XLIX.	RETURN to FORD	312
L.	DESIRE OVERCOMES OBJECTIONS	316
LI.	LIFE CHANGES AGAIN	318
LII.	IF IT'S NOT IMMORAL or ILLEGAL DO IT AGAIN	320
LIII.	THE LOSS THAT ALL MUST ENDURE	327
LIV.	ADJUSTING THE AIRPLANE to COMPLY	331
LV.	A LESSON LEARNED	333
LVI.	THE NATIONAL CHAMPIONSHIP	336
LVII.	A CASUAL ACQUAINTANCE	340
LVIII.	HOLD FAST to DREAMS	343
LIX.	1987, OPPORTUNITY DENIED	348
LX.	THE OTHER LOSS	352
LXI.	1988, HE HITS HIS STRIDE	354
LXII.	CONFIDENCE GROWS	359
LXIII.	THE NEW OBJECTIVES IN SIGHT	363
LXIV.	JUNE 17, 1990	370
LXV.	CONTINUED COMPETITION	373
LXVI.	AIR SHOW FOR THE SOARING CLUB	377
LXVII.	A NEW START in LIFE	379
LXVIII.	NANCY & DAVE	382
LXIX.	CAREER'S END	386
	EPILOGUE – PART VII	390

EPILOGUE – PART XXII 391
EPILOGUE – PART XXV 392
EPILOGUE – PART XXVIII 393
EPILOGUE—PART XXX 394
EPILOGUE – PART XXXI .395
EPILOGUE – PART XXXII. 396
EPILOGUE – PART XXXIV 397
EPILOGUE – PART XLIII. .398
EPILOGUE – PART LXIV. .400
EPILOGUE – PART LXV 401
AFTERWORD. 402

IMAGES

STANDARD J-1 (courtesy Experimental Aircraft Association Library)......3

CURTISS JN4D "JENNIE" with OX-5 engine. (courtesy Experimental Aircraft Association Library)......4

COMMANDAIRE (courtesy Experimental Aircraft Association Library)......10

The only advertisement saved from Mac's barnstorming career (author's collection)......11

The last certificate issued to "Mac" courtesy FAA files......13

Jewel & Mac in 1934 (author's collection)......17

Curtiss-Wright Junior (courtesy Experimental Aircraft Association Library)......20

HEATH PARASOL with Henderson motorcycle engine(courtesy Experimental Aircraft Association Library)......24

WACO –10 biplane (courtesy Experimental Aircraft Association Library)......25

TAYLORCRAFT BC-12 (courtesy Experimental Aircraft Association Library)......30

David & Jimmy in 1943 or '44 (author's collection)......38

BRUNER-WINKLE "BIRD" with 100 H.P. Kinner engine(courtesy Experimental Aircraft Association Library)......43

PIPER J-3 CUB (courtesy Experimental Aircraft Association Library)......45

AERONCA 7AC "CHAMPION" (courtesy Experimental Aircraft Association Library)......45

Luscombe 8-A (Shown with "rag wings". Mac's had the all metal wings) (courtesy Experimental Aircraft Association Library)......48

The Toliver House (author's collection) 60

Jimmy's 1923 Model T with David driving, Evelyn Cannon, Nell Cannon & Jan Hendrix in the back seat. The toddler is a cousin of the Cannon twins. (author's collection) 90

Case magneto from an 85 H.P. Continental airplane engine installed on the '23 Model T Ford engine by "Mac". (author's collection) .90

Model "A" Ford engine Note the copper strips between the distributor and spark plugs (author's collection) 92

'50 Plymouth with '49 front end (author's collection) 133

Hot Rod project with '56 Ford engine, '40 Ford running gear on Model T frame (author's collection) 133

Jan Hendrix's Renault in the creek eight miles north of Evergreen on U.S. Highway 31, July 31, 1957 (author's collection) 138

Aeronca "Defender" in the hangar on the site now occupied by the "Big Red" hangar (author's collection) .147

Preparing to fly after work in April, 1959. Bobby Stewart is assisting. (author's collection) 147

Pre-flight inspection being performed by Dave McKenzie on the "Defender" (author's collection) 148

"Mac" McKenzie giving rides to CAP Cadets in the L-16 in 1961. (author's collection) 172

CAP Captain Dave McKenzie and the L-16 at Tuscaloosa, Alabama, May, 1961. (author's collection) 172

Cessna 172 flown by Chuck Prince & the Hot-Rod owned by Eddie Edeker & driven by Dave McKenzie in the air show at Middleton Field on October 1, 1961. (author's collection) 174

The opening act in the Thrasher Brother's Air Circus with Grady performing in the "Twin Ercoupe" circling his brother, Bud making the "American Flag Parachute Jump" (photo from a postcard given to the author by Grady Thrasher) 183

Photo of the "Cortez" motor home copied from the sales brochure. (author's collection) 193

"Sylvester" and "Stinker" on the ramp for fueling at Greenwood, South Carolina (author's collection) 226

Steve Wittman's given name was, Sylvester. (author's collection) ..227

Great Lakes upper wings under construction in David's basement. (author's collection) 232

Mac performing the final welding on the Great Lakes fuselage. (author's collection) 233

Great Lakes fuselage ready for return to Michigan from Alabama (author's collection) 233

Great Lakes N-88-SK enroute to Grosse Ile, Michigan December 3, 1977 (photo from author's collection taken by Gail & Bob Jackson) 264

1973 Ford Galaxie after collision with locomotive (author's collection) 310

McKenzie's Landing, Howell, Michigan (photo author's collection) 385

FORWORD

BOOK ENDORSEMENTS:

I have known Dave McKenzie for close to fifty years and have always envied his talents as a pilot, a flight instructor and builder of a beautiful Great Lakes aerobatic biplane. But after reading a copy of "The Spirit's Journey" I have seen another side of Dave that is revealed in this book. Dave is a man of determination and dedication in overcoming a challenge. A man who will not take "No" for an answer. After a near-fatal airplane accident Dave's doctors told him he would never walk again. But he did! They told him he would never return to his job at Ford Motor Company. But he did! And they told him he would never fly an airplane again. But he did! Not only did he fly an airplane again but, he participated in a number of aerobatic contests following his recovery! "The Spirit's Journey" describes Dave's lifelong involvement in aviation with many interesting and humorous anecdotes and is a "must-read" for most any aviation enthusiast. But perhaps even more importantly this book describes how Dave was able to overcome the dire predictions of his doctors and therapists and how he proved them wrong. For that reason alone I would highly recommend this book to any individual who has suffered severe trauma. Dave has shown how with determination and perseverance, with tenacity and dedication, and with pure "guts" he was able to resume a normal life. This book could serve as an inspiration to those individuals who are facing similar challenges in their lives.

Bob Pauley: Private and glider pilot, aviation photographer and author.

Dave McKenzie

Having lived in Evergreen, Alabama all my youth, I knew the McKenzie family well. While I am sure that everything in this book is true, I also know that there are many, many untold stories that should be said; tales of steep hills and old bicycles and old pup tents. I know where a lot of bodies are buried. These stories should also be told.

In writing this testimonial for Dave, something came to mind that I had not thought of in years. There is a story buried in this manuscript of when Mac McKenzie flew my mother and myself for my first airplane ride. After the flight (I must have been three or four years old), Dave and I sat in the airplane and talked. He explained all the controls, instruments; and how they related to flying an airplane. I was hooked for life. He is responsible for my start in what turned out to be a very great and successful career in aviation. Thanks, Dave!

The remembrance of your growing up will return time and time again as you read of the McKenzie family. And, you will laugh often and hard. Enjoy!

Captain John D. Patten
Delta Airlines, Retired

There's little doubt in my mind, there's something we learn about ourselves, from the people that enter our lives.

I met Dave McKenzie through our local EAA Chapter when I was young, and over the years, I've learned so much about myself from him, his family, friends, and our experiences. Dave was my flight instructor when I passed my private and instrument pilot check rides. He stuck with me when I had a forced landing due to my own neglect. During my High School years, I worked evenings and weekends with Dave, building his Great Lakes Biplane, gaining the knowledge and confidence to build my own plane.

Through my friendship with Dave I've been fortunate enough to meet people in all walks of life that have given me perspective of the world, aviation, and of myself. It wasn't until Dave's untimely accident though, that I awoke to the true frailty, depth, and importance of the lives we live. It gave new perspective to all of the experiences in my past and new outlook to the experiences of my future.

After reflecting on Dave McKenzie's determination to succeed, courage, and will to overcome the adversities as told in his book, I hope you gain insight to your life's perspective, its experiences, and that you become the person you truly wish to be.

John O. Maxfield, Corporate Pilot

INTRODUCTION

The beginning of all of our lives may have begun in the infinite past of time and space in a sphere so infinitely small that it would have been impossible to see or feel. Compressed into that infinitely small sphere would have been the nucleus of all the elements that compose the bodies of all things and living creatures that would appear in the future. Perhaps there was an entity composed of energy called, "Spirit" that selected the mixture of elements that became "homo-sapiens" to be the vehicle to transport it through time and space. Perhaps there was an explosion at that time and place that is expanding towards the opposite infinity of the beginning. Perhaps it has already reached that infinity and is now shrinking backwards toward its origin in the original infinity, the beginning.

Somewhere between those two unimaginable limits of infinity there is the time and place where our ancestors, descendants, and we existed and lived and were the vehicles that transported THE SPIRIT of life on its JOURNEY from infinity at the beginning to infinity at the end. Blaney Jewel Currence and George Dewey McKenzie existed, lived, and produced descendants in their own time and place between those two extremes of infinity.

Kenneth McKenzie was a Scotchman who immigrated to North America through Wilmington, North Carolina around 1780 and settled in the Pee Dee River area of South Carolina where he married Miss Mary McLaurin. One of their sons, Duncan and his wife, Barbara McLaurin moved westward into that part of the country now known as southern Mississippi and settled into farming near the community that became known as Bay Springs. One of their sons, John A. became a sergeant in the Confederate army, was captured at the surrender of Vicksburg, paroled, forced to rejoin the army, and was captured a second time at the Battle of Nashville in December, 1864. He died a Confederate POW in Camp Chase, Ohio on January 30, 1865. He had fathered two daughters, three sons, Daniel, John Duncan, and Allen Lee McKenzie.

John Duncan McKenzie fathered two children by his first wife who died of unknown causes, three children by his second wife who died from pneumonia, and his third wife, Ollie English delivered a baby on February 26, 1906 that they named George Dewey McKenzie.

Dewey McKenzie was introduced to aviation in 1926 and made it one of his, "passionate loves" for the next fifty years. His eldest son, David inherited and nurtured the same passionate love of aviation that became one of his two principle interests for most of his life.

"The Spirit's Journey" is the review of seventy years of David's life that had both the love of aviation and automobiles that led to coexistent careers as an automobile designer, an airplane pilot, a flight instructor, an aerobatic pilot, and a family head.

PART I

GEORGE DEWEY MCKENZIE

The family of the youngest child of John Duncan McKenzie and Ollie English who was twelve or thirteen years younger than John included the two older brothers and three older sisters by his father's first and second wives and all of them grew up on West 10th Street in Laurel, Mississippi. Ollie and John did have one other child together that died in infancy prior to Dewey's birth. That might explain why Dewey was raised in a manner that later resulted in his being described as a "spoiled brat".

George Dewey McKenzie quit high school when he was sixteen years old and entered the work force probably to satisfy his interest in vehicles that were self propelled. He had been driving automobiles since being twelve years old and it isn't known when he started riding motorcycles which he called, "motorsickles". Incidentally, when challenged about the correct pronunciation of the vehicle's name he'd ask, "Did you give your daughter/son a bicycle or a bysickle?"

His additional schooling was both the "hard knock" variety and some formal training. The night school mechanic's class instructor who taught him how to, "pour a bearing" in an engine block was Ray Harroun, the winner of the first Indianapolis 500 automobile race in 1911 driving a Marmon Wasp automobile.

Sometime in 1925 Dewey and his father, John had a familial disagreement and John evicted his son from the home. With no place to live he left Laurel and joined his next older brother who was the maintenance foreman at the Royster Fertilizer Company plant in Savannah, Georgia.

His brother had met, fallen in love with, and married a nurse he'd met in France during World War I who had a daughter slightly younger than Dewey from a previous marriage. The, "French Nurse's" daughter and Dewey married.

One afternoon Dewey's older brother arrived home after working late and found his supervisor in bed with his wife, the "French nurse". He responded by fatally shooting his supervisor and the ensuing legal battle for his freedom which his father helped support financially was monetarily devastating to both families. That was probably the source of the pressures that ultimately led to the dissolution of the union of the sixteen year old, Alma and the nineteen year old, Dewey.

After moving to Miami, Florida he was a mechanic at the, "Yellow Cab Company" for a short time. Then, Dewey moved on to Meridian, Mississippi probably because his oldest brother was there and found employment as an automobile mechanic. That led to his being called to the local airport one Sunday morning in 1926 to repair the magneto on an airplane. The compensation he was given for the repair was a ride in the airplane.

PART II

THE BUG BIT

Today on the east side of Meridian U.S. Highway 11 splits into the, "bypass" that goes south around the city and the "old road" leads straight ahead into the city. That intersection is at the approximate location of the northeast corner of Meridian's first airport, "Bonita Field" which was the base of operations of the, "Key Brothers' Flying Service" when they went into business and it was the airport where Dewey first touched an airplane. He went there to take flying lessons and soloed in six hours flying a Standard J-1 biplane He later said that "when he soloed and taxied back to the office the operator ran out and said, "Don't shut the engine down. I have a passenger for you to take for a ride." He gave the man his ride and was given his first student to teach to fly when he had thirty hours total flying time in the Standard J-1 airplane.

STANDARD J-1 (courtesy Experimental Aircraft Association Library)

Dave McKenzie

A gentleman in Americus, Georgia, John Wyke had purchased an unknown number of Curtiss JN4-D biplanes from the U.S. Government on the surplus market after World War I concluded and then offered those airplanes for sale to civilians. The airplane is commonly known as the, "Jenny." Charles Lindbergh purchased his first airplane there. Dewey, who may have begun to use the nickname, "Mac" by that time has said, "You could go to Americus, give the man a hundred bucks and he'd help you uncrate a, "Jenny" of your choice and assist you in assembling it. You'd take the valve action off the engine and replace it with your own, "Miller" valve action and you had a brand new airplane to start the barnstorming season with. An OX-5 engine had a life expectancy of about a hundred hours provided you took real good care of it and by that time the airplane would be so "beat up" that it was junk. One hundred hours was about all the barnstorming you could do in a season so, you'd remove the Miller valve action, junk the airplane, winter in somewhere, and the next spring you'd catch a train and carrying your "Miller" valve action go back to Americus."

The Miller valve action had one shaft per cylinder bank that all of the valve rocker arms pivoted on and the zerk fitting for greasing the assembly was located on the end of the shaft. The standard valve operating mechanism on the OX-5 had a grease fitting on each rocker arm. That amounts to two valves per cylinder multiplied by eight cylinders equaling sixteen rockers to be lubricated which "Mac" said, "Had to be done at the end of each day so that you'd be ready for the next day's flying". It is believed that Mac wore out three of the, "Jennies".

CURTISS JN4D "JENNIE" with OX-5 engine.
(courtesy Experimental Aircraft Association Library)

At one time, Mac was a smoker. That is he was addicted to the inhaling of smoke from burning tobacco wrapped in paper. It was a habit that he quit in 1935 or '36 and passionately hated for the rest of his life. He claimed that the barnstormer's other duty at the end of the day required that he lay his "Taylor Cigarette Rolling Machine" on the top surface of a lower wing along with the required supply of cigarette papers and tobacco supply and roll the next day's supply of, "smokes". After all, no aviator should appear to be so poor that he had to "roll his own" and if asked what brand he smoked his response was, "Taylormade."

Another standard joke was to respond to the question, "Are you married?" by answering, "Hold on, let me check the map".

The inquirer would always ask, "What's that have to do with it"?

"Depends on how far I'm from home"

There was some "rebel" in his psyche. He refused to "log" that is maintain a record of his flying time therefore, it isn't known when he flew what, where, or how long it required. He did mention owning and/or flying the Detroit Parks, Lincoln-Paige, Curtis Robin, Waco 10, Waco GXE, Curtis Jenny, American Eagle, Swallow, Commandaire, Travelaire, Standard J-1, Aeronca C-3, Aeronca K, Ford Tri-motor, Fairchild 24, PT-19, PT-26, Stinson 108, Stinson 10, Taylorcraft, Piper/Taylor E-2, Piper J-3, Luscombe 8-A, Brunner-Winkle Bird, Meyers OTW, Waco UPF-7, Ercoupe, Great Lakes, Cessna 150, 172, 170, Barling, and the list goes on seemingly endlessly with at least one story accompanying each one of them.

By 1930 Dewey (the populace in Laurel never knew him as "Mac") was one of the pilots at "Stump Field" located at Laurel, Mississippi along with Max Holerfield, Pat Mulloy, Alton Hesler, and James Daniels. Dewey McKenzie and Max Holerfield had a contract with the Mississippi State Fair touring from town to town in the state performing one-week stands. They were the, "Flying Circus" part of the fair. The fair did not list them on the payroll and only supported them by listing them in advertisements and usually making arrangements with property owners close to the Fair Grounds in each community for the "Air Circus" to use their cow pasture, cotton field, or whatever for an airport during the fair. It is believed that this arrangement was consummated for three consecutive years. The "modus-operandi" was to perform an air show each day to attract a crowd that they would attempt to sell rides to after the exhibition flights were completed. That was their only source of income and they have said that some days they sold absolutely no rides and then on others they couldn't haul all the passengers.

There is no way to know the date and location of the attempt by Max and Pat to announce the arrival of the State Fair and its accompanying troupe of daring, appealing, handsome, financially blessed, suave, charming, accomplished, and worldly wise aviators with their sophisticated modern airplanes powered by proven, reliable, dependable, and powerful engines but, they had arrived at one community and as usual one airplane was making their arrival noticeable to the residents of the community by flying low over town.

Max was flying the airplane and Pat was in the front seat, their usual positions, and Dewey waited in the cow pasture that was going to be the, "airport" for the week for the results of the inaugural flight over the town.

After arriving over the town Pat climbed out of the front seat and worked his way out to the interplane struts between the lower and upper wings as Max flew around above the community. As they continued the operation, Pat climbed down to the spreader bar between the wheels and sat down. It should be noted that the performance was being executed prior to the installation of zippers on coveralls. (Dewey, for all of his life called them, "rompers".) Pat was sitting behind the propeller and the blast of air flowing rearward from it combined with the air flow generated by the forward movement of the airplane was more airflow than the buttons on the coveralls could tolerate. The "rompers" opened down the front and the air blast inflated them like a balloon. Every time Pat succeeded in buttoning one fastener and released it to button the next one the button that was previously closed would blow open. He couldn't climb back up to the lower wing and then into the front cockpit with his "rompers" inflated and creating the aerodynamic drag that they would generate so, he proceeded to remove them.

Of course, Max couldn't see Pat on the landing gear below the fuselage from his position in the rear seat so; he didn't realize that Pat was in trouble. He just wondered why Pat was remaining down there so long. He could see people on the ground coming out of stores, stopping cars, exiting their homes, looking up, waving, and exchanging stares with each other. He thought, "Business is going to be good".

Pat had removed his coveralls and remained seated on the spreader bar while trying to figure out how to hold onto his, "rompers" while climbing back to the cockpit dressed only in his drawers, and socks. Max had continued to fly around over the town with the nearly nude wing walker on the landing gear until Pat finally succeeded in climbing back to the lower wing.

It is not known what was thought, reported, or to what authorities, by curious onlookers, representatives of the local churches, the local gendarmes, or potential paying passengers but, no one or anything was injured physically.

Max was illiterate and gave the impression that he was not sophisticated or refined but, barnstorming pilots didn't have much contact with their passengers. Max raced "motorsickles" professionally until he was fifty-five years old and was never seriously injured.

Pat Mulloy on the other hand was from an established family who were also financially successful and never attempted to make aviation his career and yet, he was very accomplished and known internationally in later years for his achievements flying sailplanes.

Alton Hesler worked as a mechanic in various automobile and truck garages around Laurel, he became an accomplished pilot and aircraft mechanic, and was later the co-owner of Hesler-Noble Flying Service in Laurel at the airport now serving the community which grew from the original "Stump Field" and is now named, "Hesler-Noble Airport".

James Daniels was a professional pilot for the duration of his life who participated in some short barnstorming trips with Dewey and is probably best remembered for landing an airplane on Canal Street in New Orleans, Louisiana.

James, Dewey, and other pilots in the Southern Mississippi and Louisiana area had met at the new airport beside Lake Pontchartrain and north of the city of New Orleans to participate in the dedication and opening of the airport named, "Shushan Field" on February 10, 1934. Shushan Field is now named, New Orleans Lakefront. Of course, the objective James, Dewey, and the rest of them had was to haul passengers for their financial benefit.

James Daniels had apparently flown down from Laurel sometime during the preceding week because he had sold a sight seeing flight over New Orleans to a customer prior to the arrival of the other pilots.

During the flight he experienced an engine failure and selected Canal Street as the site for the ensuing forced landing which was successful. He got out of the airplane, opened the engine cowling and was making the necessary repair when the city officials arrived, offered their sympathy and assistance, and when he had completed the repair, blocked traffic for his takeoff and return to the airport.

Not much time was required for the story to circulate around the airport and as should be expected there were those on the airport who questioned the possibility of such an event ever occurring. The world wide financial depression was at its worst and James Daniels recognized the opportunity for a financial realignment. Specifically, "Change their money into mine". Wagers were offered on the feasibility of landing on Canal Street.

After the airport dedication was completed James Daniels took off, flew over downtown New Orleans, closed the throttle, landed on Canal Street, got out, and opened the cowling.

The city officials and gendarmes, after noting that this was the second time in less than a week that this pilot landed on Canal Street, refused to give his assertion of it being an emergency very much credibility. In their generosity they offered him overnight lodging in the local calaboose and insisted that he accept the offer of their hospitality however, once they got him in it they did not seem to appreciate him contaminating their calaboose. He was soon released and they blocked the traffic as he flew out after intently listening to their admonitions to stop landing on Canal Street.

One of the other airplanes flown at the airport dedication that weekend was a Curtiss Robin owned by the writer, William Faulkner of Oxford, Mississippi. That airplane was used in 1933, '34, and '35 as the refueling plane for the Key brother's endurance flights at Meridian, Mississippi. The pilot for the, "refueler/service" plane was Jim Keaton who later became a Captain for American Airlines and lived in Foley, Alabama after his retirement.

William Faulkner wrote a novel based on that airport dedication entitled, "Pylon" that the movie released in 1958 named "The Tarnished

Angels" starring Dorothy Malone, Robert Stack, and Rock Hudson was based upon.

Prior to the outbreak of World War II James Daniels went to Canada, joined the Royal Canadian Air Force, was stationed in England, ferried aircraft untial, in self defense, he shot down a German fighter. He was then transferred to combat duty. After World War II he returned to Mississippi and it is believed that he was overcome by the vapors coming from the dust hopper of a "duster" that he was flying and was fatally injured in the resulting accident.

While operating as the South Mississippi State Air Circus Dewey and Max once spent a week at Bassfield, Mississippi and on the final day at that location a prospective passenger arrived early in the morning so, Max was taking him for a ride in weather conditions that were nearly perfect and with no wind. Of course, there was no agreement between Dewey and Max on the selection of the, "calm wind runway" so, Dewey watched Max leave taking off toward the end of the runway that had the power lines crossing it. He expected Max to approach for landing from the other end in the calm conditions so that he didn't have to cross the obstruction caused by the power line.

Dewey has said, "If you ever get lost in an airplane simply select a power line and follow it. It'll eventually lead you past the end of a runway. Also, the only pilot who's never been lost is the one who's never gone anywhere."

Another, "paying passenger" arrived and Dewey, "converted his money from, "his to mine", loaded him into the airplane, a "Command-Aire" and took off. After his take off another passenger appeared and Max departed taking off in the opposite direction from the one Dewey had selected.

Of the two flights, Dewey's was the first to return so, he approached from the obstruction free end and just prior to touch down he sighted the other airplane approaching from the "wrong end". He held the airplane low to allow the other airplane to pass above his, misjudged the clearance to the power line, hit it, broke the propeller, stopped the engine, and eliminated the offensive power line. The next obstacle in his path of flight was a house, which he then maneuvered around and found himself faced with a series of terraces crossing his direction of flight as he followed the hillside onward. He was attempting to stretch the glide to level ground beyond the terraces when the airplane finally stalled, "dropped in", struck a terrace, and nosed over onto its back (upside down).

Dewey grasped the fire extinguisher in one hand, released his seat belt with the other, and fell out of the inverted airplane. He complained later about the fire extinguisher mounting bracket causing a cut on one finger.

Gasoline was leaking out of the inverted airplane and he could hear the passenger kicking or beating something in the front seat so, he got up, reached into the front seat area, and released the passenger's seat belt who promptly fell onto the bottom surface of the upper wing center section. The passenger's legs were bleeding below the knees where he had injured them by kicking against the corner of the fuel tank located behind the front instrument panel. There were no other injuries.

COMMANDAIRE (courtesy Experimental Aircraft Association Library)

The airplane was disassembled and returned to Laurel where it was repaired at, "Stump Field" near Laurel. Housewives who resided in the neighborhood of the airport hung their laundry out to dry in the evening so that the aviators would not convert their bed sheets into aircraft covering material. The world wide financial depression was at its peak and aircraft parts and supplies were still expensive.

The Spirit's Journey

THURSDAY, SEPTEMBER 28, 1933   THE PRENTISS HEADLIGHT   Jefferson Davis County, Mississippi

> **AEROPLANE CIRCUS**
> ONE WHOLE WEEK
> **STARTING SUNDAY, OCT. 1st TO OCT. 7th**
> 2 Aeroplanes,   2 U.S. Licensed Pilots
> Safe Landing Field
> ENJOY A REAL THRILL
> Rates for Rides **75¢ to $1.50**
>
> Flights Arranged to Suit Your Convenience
> STUNTS OVER FIELD EVERY DAY
> **SEE** The Daredevil Pilots Flirt with Death
> Loop the Loop, Fly Upside Down, Walk the wings
> —Take the Dangerous Nose Dive—Change from
> one cockpit to another—and do every thing that
> a pilot could possibly do with an Aeroplane.
>
> **REMEMBER:** See the Death Dealing Exhibition over the Airport Each Day.
> **NOTICE TO THE PUBLIC.**
> One of these Ships is a new Lincoln Page Powered with a 480 H. P. O X 5 Motor. The other is a Comandaire Powered with a 480 H. P. Curtis Motor. The Comandaire is known to be the Safest Ship that flys in the air to-day. You take no chance in making your first flight in either of these ships. They are as safe as your automobile. Come out and get a real thrill for the first time.
>
> For Further Information See:
> **N. D. McLEAN, Bassfield, Miss.**
> Airport Located One-Fourth Mile From Bassfield on Highway 42.

The only advertisement saved from Mac's barnstorming career (author's collection)

Two of Dewey's older sisters had moved to Tylerton, Mississippi. One had married Dewey Collins who was related to a former Governor of the state and the other had married a sawmill owner and moved there with him and their children. Dewey would ride his, "motorsickle" to Tylerton regularly during the, "off-seasons" for visits.

Beside a road, which was on a hill near one of his sisters' home, was a gravel pit, which he soon dug an opening to for access from the gravel road providing suitable ingress/egress for his, "motorsickle". He would ride up or down the hill in the road and when he reached the gravel pit he'd turn into it and roar around the wall with the bike perpendicular to the wall held onto the wall by centrifugal force just as has been performed in carnivals at the exhibit called the, "motordrome".

Dewey's principle source of amusement in Tylerton was to invite someone to join him on the bike for a brief, "pleasure" ride and then take him or her around the gravel pit without announcing his intentions. He was usually the only one to derive any pleasure from the ride. Like everyone else, he exhibited some characteristics in his psyche that were not completely acceptable to some of the people observing them. One of those traits was to adopt deportment that displayed a lack of etiquette and mannerisms that would be associated with ignorance, crudeness, and lack of sophistication. He really was none of those things but sometimes he'd assume the behavior to deliberately embarrass some who were emotionally close to him. Perhaps his mannerisms were sometimes an imitation of Hughey and Dewey Long of the Louisiana political dynasty. If it, "pissed you off" he loved it. Overall though, he was no one's enemy and those who did annoy him had to work hard to do it.

In 1926 the United States Government decided it was necessary to regulate and control aviation for the benefit and safety of the public. The Department of Commerce started testing pilots and mechanics, inspecting aircraft, issuing licenses and certificates, and composing standards and requirements for the performance of everything, and everybody involved in aviation. Of course, those things were done by decree of that collection of, "bureaurat" conceiving money eaters called, "Congress".

About 1930 Dewey was based in Laurel and it isn't known what type airplane he owned at that time. It is a fact that he did not have a Pilot's License. None of the other pilots operating from, "Stump Field" had a license either.

What good is a piece of paper? It won't provide the skill, talent, knowledge, ability, or anything else. It just provides a job for conceited, self-serving, officious "bureaurats". At least that's what one might be led to believe was the opinion of the barnstormers around, "Stump Field". They owned, flew, taught each other, and maintained their airplanes with no approval from anyone or any organization.

An, "Inspector" from the Department of Commerce arrived at, "Stump Field" and directed the pilots, mechanics, and anyone else interested in flying to line up in front of a table he commandeered, copied their names, addresses, asked what function they served, which airplane they flew, etc., and then left.

A couple of months later the U.S. Mail carrier delivered a small wooden jewel box wrapped in a red, white, and blue ribbon which when opened revealed on its blue velvet lining a, "Transport Pilot" certificate numbered 24031. In later years it became a Commercial Pilot Certificate. Pilot certificates were numbered in chronological order so; it indicates that there were only 24,030 pilots licensed before George Dewey McKenzie in the United States.

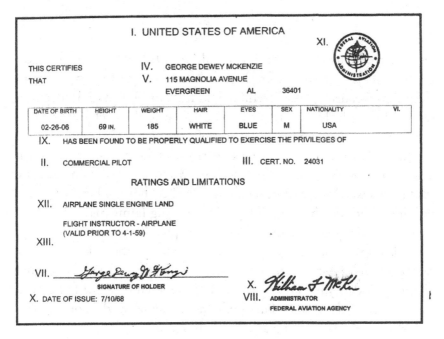

The last certificate issued to "Mac"
courtesy FAA files

Mac became acquainted with Monroe Sanders who had grown up in Brooklyn, Alabama. Monroe asked Virginia Hunnicut of Cuba, Alabama for a date one evening in the spring of 1932 and she accepted with the provision that he bring a friend along to accompany her friend who lived down the road.

PART III

JEWELL CURRENCE

Jewel Currence had begun her life on July 13, 1912 and was the youngest of the four sisters and older than only one of the four brothers in the family headed by Anne Miller, and husband Julien Currence that resided in the little community called Cuba on U.S. Highway 11 in Alabama just twenty seven miles east of Meridian, Mississippi.

Jude Miller, Anne's father was native to Choctaw County and it isn't known where Julien Currence's family originated from because his parents were fatally shot by a guest in their home when he was an infant and he was then raised by a neighboring family named, "Hall" that never changed his name or legally adopted him.

It is known that Jewel didn't always maintain the strictest control of the emotion identified and described as anger. That personality characteristic may have been genetic and inherited from her father who was known for periodic outbursts of temper. Her father died of coronary failure at a relatively young age, his early sixties.

The periodic exhibition of the verbal attempt to remedy frustration, verbally relieve tension caused by uncooperative devices and people, and also be a substitute for any medically prescribed sedative inspired Jewel's sons to ask their maternal grandmother why she had allowed her daughter to, "cuss".

"Granny", as she was known to all of her grandchildren described how she had broken the butter paddle on, "Jewel's little behind" when she was six years old and still couldn't get her to desist in the use of profanity which she probably learned from her father, Julien. Granny said, " She cussed every time I hit her with the paddle so, I just gave up."

Jewel had graduated from High School in Cuba with a good academic record, was a pretty girl, had all of the social graces, and became a beautiful woman, intelligent, refined, industrious, and later, a good mother to her two sons. She was known as a meticulous housekeeper and became an antique furniture collector who was known by dealers from Atlanta to New Orleans and from Pensacola to Birmingham.

Jewel Currence was engaged to marry a young man from Whitfield, Alabama who was attending college near Chicago, Illinois and hoped to become an electrical engineer. He contracted pneumonia while in school and died but, Jewel never forgot his mother, Mrs. Maggard or his Aunt, Miss Belle Flowers who were referred to as Mrs. Maggard and Aunt Belle for the rest of their lives by Jewel's family. Jewel's sons vividly remembered spending several nights in the two storied house built before the 1850's in Whitfield of squared logs with the, "dog trot" splitting the lower story running from the front porch to the rear one.

Monroe Sanders and Mac met the girls at Virginia Hunnicut's home in Cuba and went for a drive in the Essex roadster owned by Monroe followed by snacks, beverages, and small talk back at Virginia's home. There was no discussion of any future dates but, Mac did learn that Jewel was going to visit friends in Whitfield that coming weekend so, he decided to, "barnstorm" the Whitfield area that weekend.

Mac has described the typical/traditional line of spectators at his barnstorming, "events" as rows of people standing beside the fence watching the airplane land, taxi up, turn around, unload the passengers, reload, and takeoff again even though their faces became coated with a mixture of perspiration and dust blown up by the airplane's slipstream when it turned around to the point that their eyeballs and teeth would contrast with the color of their faces. "They looked like, "Minstrel Show" black face performers. Of course, most of 'em were hopin' to see an airplane *crack up*."

Near sundown that Sunday afternoon in Whitfield Mac addressed the crowd saying, "That's all for today, folks. I want to take my girl friend for the last ride of the day."

Jewel got in the front seat of the, "Command-Aire" open cockpit biplane as Mac held the small door on the side of the fuselage open, then closed it after fastening the seat belt. He, "hand propped" to start the engine running, got in the pilot's seat, and took-off to very smooth flying conditions. The sun was, "on the horizon". The light breeze died and the earth had started to cool which eliminated the generation of heated columns of air rising that would then cool when reaching higher altitudes and descend to make contact with the warm earth and start the entire process again. Except for the steady, smooth sound made by the engine turning the propeller it was like sitting in the living room at home except for the panoramic view of the planet slowly drifting by and in so doing revealing all of its secrets to view. Whitfield is just west of the Tombigbee river and approximately fifteen miles south of the small city named Demopolis which is the second oldest settlement in Alabama, second in age only to Mobile and both communities had been established by the French settlers in the early 1700's. They were flying over a sparsely settled agrarian area with the longest European history in Alabama.

Suddenly Mac's vision toward the front was blocked by the appearance of, "skirts, slips, bloomers, feet, legs, arms, hands, and hair" as he described it. His own seat belt caused pain as the entire airplane suddenly, violently dropped, and his feet rose upward from the rudder pedals with his legs striking the lower edge of the instrument panel. Then, just as abruptly as it had begun it's descent and the airplane with it behaving just as a molecule of that mixture of gasses named air behaved the downward flow of air ceased. Mac was slammed down into his seat. He later said, "Jewel had every hand, foot, both legs, and maybe every tooth wrapped around the struts and wires." She "climbed back down into the front seat".

Neither Mac nor anyone else in his generation of pilots had ever heard of a phenomenon called, "Clear Air Turbulence, (CAT)" which is what they had encountered. When the airplane had dropped out from under them the upper wing center section struck Jewel on her head, the back of her neck, and then her back. It was the upper wing center section that saved her from going overboard and falling into, "infinity".

Mac flew the airplane back to the little pasture they had departed from, landed, and Jewel went home with Mrs. Maggard and Aunt Belle. That was their second, "date". Many years later when questioned about that experience Jewel would only say in a dulcet tone with her southern accent, "Well, that's about right. It happened."

Mac would describe the event as it is written and then add, "I thought the damned fool had her seat belt fastened."

The relationship was continued, grew, became warm, and their acquaintances knew that Mac had found something in addition to airplanes that he loved.

It is known that Mac always had at least one "motorsickle" for his personal transportation and he undoubtedly owned automobiles, too. There is no report or rumor that indicates Jewel was ever on a, "motorsickle." She did visit his relatives in Tylerton several times and it was on one of those visits that Mac was sleeping late one morning, an "activity" that he developed, perfected, and regularly practiced for all of his life while the girls, Jewel, Florence, and "Buster" (Florence and "Buster" were his nieces.) drank coffee or hot tea at the kitchen table when one of the girls said to Jewel, "He obviously really loves you."

Jewel responded, "He'd love me twice as much if I made airplanes instead of babies."

Jewel & Mac in 1934 (author's collection)

Because of his interest in, "motorsickles" Mac was acquainted with a, "motorsickle" cop in Meridian who had the last name, "Rogers". He was the brother of the folk music performer and writer called, "the Yodeling Brakeman", Jimmy Rogers who wrote and recorded many country and folk songs before he succumbed to tuberculosis. Jimmy Rogers would occasionally when passing through Cuba on Highway 11 see Mac's bike parked in the Currence's front yard and the family sitting on the front porch. He'd stop, visit for a time, and during those visits he'd usually borrow a guitar and do an impromptu performance. Obviously, he enjoyed what he did.

Jewel and Mac would occasionally visit one of the two airports that existed in Meridian by that time on weekends. Dr. Key had been influential in the acquisition of property and the construction of the Municipal airport on the southwest side of town, that still exists, and his sons, Fred and Algene were the operators and it is now named, "Key Field".

Jewel later admitted that she enjoyed those visits even though in later years she seemed to hate airports that she said were, "the coldest places in winter and hottest in summer and always inhabited by bums". Yes, they did meet some of aviation's big names during that period and they did indeed have opinions of them. One of those celebrities was Amelia Earhart when she was to appear at Meridian in an airshow.

Miss Earhart was supposed to fly an autogiro in the airshow but, enroute to Meridian she, "cracked it up" and had to share a, "celebrity spectator box" with, among others, Jewel Currence and Mac McKenzie. Jewel's observation was, "She's really not pretty, not even attractive. I thought she was uuuuglyyy." and Mac said, "She can't fly worth a damn. She never flew anything that she didn't tear up."

Mac knew Jimmy Wedell and Harry Williams, the men responsible for the famous Wedell-Williams racing planes from Patterson, Louisiana very well and also Jimmy Wedell's brother, Walter prior to his death in a training accident. Walter was the instructor who had a student stall and spin in southern Mississippi with fatal results. Another of his acquaintances was Roscoe Turner who was another native Mississippian. He has said that the mother of Katherine, Marjorie, and Eddie Stinson was one of his Sunday School teachers when he was a child.

Charles Lindberg has said, "Aviation is a small fraternity and when two pilots meet they will know each other or, if not, they will both know some other pilot." Mac met Charles Lindberg once and did not speak well of him. The meeting probably occurred during 1927 when Lindbergh was executing his tour of the United States in the, "Spirit of St. Louis" after the famous crossing of the Atlantic Ocean. Mac related that he was in the hangar at Bonita Field working on the engine of one of the planes and heard a plane land. He continued to work thinking that whoever was flying the plane would get out and find their way into the hangar which is exactly what happened. He said, "Someone who seemed to be very annoyed asked, ""Where is everybody?"" and Mac replied, "Over at Key Field waiting for Lindbergh to show up." He looked around and found that the source of the voice was Charles Lindbergh who then asked, "Where's the telephone?" to which Mac replied, "In that gas station on the corner." and pointed out the hangar door. Before he could do or say anything else, Charles Lindbergh turned and walked out the door. Mac went back to work and said later that he felt that there was some arrogance in the man's demeanor that he didn't like and he never had the opportunity to learn anything different. Mr. Lindbergh had landed at "Bonita" because that was the only airport he knew existed at Meridian. It was the fourth airport he had landed on when he was beginning his career.

Oh yes, Jewel flew with him while they were dating but, there has never been anything said about it that was memorable. They married in a Judges' office in Livingston, Alabama in May 1934. Neither of them ever said much about that either.

Within days of the wedding Jewel fell ill with diphtheria and had to be hospitalized in Meridian. Mac borrowed a hearse to use for an ambulance and drove his new wife to the hospital in Meridian with no assurance that he'd ever get her back.

In the meantime he had taken employment with the Ford dealer in Livingston named, "Little Motor Company" as a mechanic and of course, the first week's salary was held and he received his first paycheck the second Saturday evening of his employment. The check was for twelve dollars and he said, "Hell, I can't live on this. I gotta' go back to flying." He quit.

Dave McKenzie

Curtiss-Wright Junior (courtesy Experimental Aircraft Association Library)

They established residence in a hotel in Demopolis where Mac began operations with a "Commandaire" that belonged to Lynwood Blackman of Laurel and he soon added a "Curtis-Wright Junior". The "Junior" was a two place, open cockpit, tandem seating high wing monoplane powered by a three cylinder Szekely (pronounce it, ("Say Kelly") radial engine produced in Holland, Michigan and mounted on the wing with the propeller behind the wing in the "pusher" configuration that was produced in the early thirties. The engine was the weak link in the airplane and it has been said that, "If there had been a good engine on the "Junior" there would never have been a market for the Piper J-3 Cub".

The most common failure of the engine was due to a weakness of the cylinder barrel to mounting flange area of the cylinders which would fail and leave the cylinder mounting flanges attached to the crankcase halves as one or more cylinders departed from the engine and airplane. The engine factory and CAA approved "fix" was to wrap a cable around the engine with a turnbuckle in it and tighten. Of course, that cable-turnbuckle assembly had to be a factory produced, CAA approved part with the appropriate price tag attached.

Alton Hesler owned one in Laurel and his was plagued with a tendency for the propeller to loosen on the tapered crankshaft. The inevitable occurred. It threw the propeller which tore through both longerons on one side of the fuselage and severed both elevator cables on that side but, did not cut a rudder cable as it departed the airplane. Hesler claimed that he was fascinated at first, watching the propeller tumble earthward and then remembered that he was faced with a forced landing. That thought refocused his attention on reality and he succeeded in making an acceptable forced landing while pressing on both rudder pedals to keep the empennage attached. The point being, the Szekely wasn't thought of as being a good engine.

The engine on Mac's "Junior" "gave up the ghost". He couldn't financially afford to purchase a replacement and "It was too good an airplane to junk." Therefore, he visited a junkyard and purchased a "Model A" Ford engine from a wrecked automobile to modify and replace the Szekely engine with.

He claimed that he wrote Ford Motor Company enquiring about their opinion of the idea of modifying the "Model A" engine and installing it on the "Junior". Their response was not encouraging and they did not offer any support for the project. It isn't known whether or not they wanted a report on the results if it was ever attempted.

He purchased an aluminum cylinder head from an, "after market" performance parts supplier, made his own aluminum oil pan, installed a magneto for ignition, designed, and machined an adapter to mount the propeller from the Szekely on the crankshaft and modified the, "Junior" engine mount to accept the Ford engine and mounted it on the airplane. He mounted the, "Model A" radiator ahead of the engine and hanging over the leading edge of the wing.

Mac never exhibited much emotion or sentiment. He never owned a camera, never kept a diary, and his thoughts about maintaining a logbook have already been described. He never kept a scrapbook either; therefore, no record exists to substantiate his activities and achievements. There were never any photographs taken of the Ford powered, "Junior" that are known to exist either but, Mac did say, "He soloed several students and hauled many passengers in the little airplane."

There was a story about a trip that he made in the, "Junior" to some community north of Birmingham to an event of some type where he was to haul passengers to entertain them and enrich him. The area that he was using for an airport was so small that the description of his operating procedure was, he'd load a passenger, take off, fly around, return and land in the opposite direction from the take off, get out of the airplane, pick up the tail and manually turn it around, reload, restart the engine, and make another revenue producing flight. The local law enforcement official observing the operation remained at the site all day awaiting an accident. He was disappointed.

A tornado struck Demopolis and rolled a tree over the "Commandaire" but, did not damage the "Junior" that was tied down beside it. Imagine the frustration that was endured by Mac as he desperately attempted to establish himself in aviation during the heart of the world wide financial depression and trying to do it in rural Alabama which does not now and never has had the financial resources to claim to offer potential for the success of such a business as that which he was trying to establish and maintain.

Jewel was fulfilling the role of the loyal, supporting, and loving housewife. There are no reports or descriptions of her interests, activities, or social groups during this period in their lives. She never attempted to seek employment outside the home and never learned to drive an automobile. Her interests, social life, and activities were like those of most feminine members of society at the time, accepted, taken for granted, and then most often forgotten by others.

Following the disastrous tornado strike Mac disassembled the "Command-Aire" and Lynwood Blackmon from Laurel retrieved it. Jewel has said that prior to that development she thought it belonged to Mac. Is it possible that she didn't care what he was doing as long as he kept bringing money home? Perhaps he felt that it was his responsibility to support them financially and she didn't need to be informed of the operation of the business. There is no doubt that there was a lack of communication between the two of them that frequently incited complaints from Jewel in later years.

Mac sold the, "Junior" to a gentleman residing near Hattiesburg, Mississippi who, while circling a friends home and attempting to return the waves from spectators on the ground pushed his arm through the propeller arc severing the arm. He must have been in the rear seat to reach the propeller arc and that would necessitate the front seat being occupied by a passenger in order for the center of gravity to be close enough to its acceptable range for the airplane to fly and be controllable. Nothing more is known or has been heard of the fate of Mac's Curtiss-Wright Junior.

It was about that time that Mac took the written examination for the rating called, "Airplane and Engine Mechanic" by the CAA. It was a subjective examination and the examiner would ask the applicant to compose a paper describing the procedure for overhauling an engine. The examiner then scored it at that time and place and that examiner refused to pass Mac because he had selected the OX-5 engine as the subject of his imaginary overhaul. The examiner claimed that it was too old an engine and did not require the skills that a, modern engine such as the "flat head", thirty seven horse power, four cylinder, horizontally opposed, air cooled Continental demanded.

Years later the examiner told Mac that he had refused to pass him because he did not know Mac and knew nothing about his experience or abilities. It didn't matter. Mac was insulted, rightly so, and vowed to never attempt that exam again. He repaired and overhauled engines and airframes for almost the rest of his life and never attempted to obtain the CAA, later FAA ratings even though he was an OUTSTANDING mechanic and welder.

The experience obviously had an effect on his outlook and attitude as evidenced by an incident that occurred at Jewel's sister's home in Cuba when the husband of one of Jewel's nieces presented the, "study guide" for the Mechanic's Rating that he was studying. Mac belittled and criticized it, often laughing at the questions resulting in embarrassment to Jewel and the others who were visiting the home. R. V. Davis became a maintenance supervisor for Delta Airlines in Atlanta, Georgia.

There are times when we allow other people's egos to hurt us but, even worse, we sometimes allow our own pride to rob us. Of course, flying was still the great objective to Mac and in his desperation to own and fly an airplane he found and obtained a small airplane that had been built from a kit produced in Niles, Michigan. It was a single place, (pilot only) open cockpit, parasol monoplane, powered by a converted four cylinder, inline, air cooled, Henderson, "motorsickle" engine, and was called a, "Heath Parasol".

It had been built with the wing span reduced, obviously in an attempt to increase the cruising speed of the little, "bird" but, with so little power available from the engine it needed the long wing that its designer, Ed Heath specified to fly and there is reason to suspect that it had been seriously damaged in a wind storm.

Mac disassembled it and sneaked the wings into the hotel room he and Jewel resided in, built more ribs, increased the length of the spars by splicing them, reassembled the wings, and covered them in the hotel room in Demopolis.

Undoubtedly there was some stress brought to bear on Jewel by that activity in their hotel room. She later said, "Mrs. - - - - - - - - -, the hotel manager would stand at the bottom of the stairs and yell, ""Mrs. McKenzie, are you cooking up there, again? I smell bananas and you're not supposed to cook or eat in that room."" Jewel never elaborated on how she responded to those inquiries or what the effect was on her personal disposition.

Linden, Alabama is ten miles south of Demopolis and it was the site of an installation operated by the Civilian Conservation Corps (CCC). Mac applied for employment with responsibility for maintenance of all the automobiles, trucks, and busses operated at the facility.

HEATH PARASOL with Henderson motorcycle engine
(courtesy Experimental Aircraft Association Library)

They moved into a duplex apartment in Linden and Mac recovered the fuselage in the front yard of that home. He then took all of the components of the little airplane to a field behind the maintenance shop at the CCC camp where he had made arrangements with the property owner to start an airport and reassembled it.

He began the takeoff run on the test flight and claimed that the engine stopped running at the moment he lifted off the ground leaving him with no option but to land, "straight ahead" into a plowed field that had the rows running perpendicular to his direction of flight.

"The wheels rolled across the first row and stopped at the second." was Mac's description of the landing. Witnesses say that the plane stood on its nose, broke the propeller, and remained in a three-point attitude with the two wheels and the nose providing the three points on the ground.

Jewel was among the spectators. Mac had no injuries so, he replaced the propeller and flew the little airplane on another day.

The Heath, of course, offered no way to financially support itself or its owner so, it was replaced by a Waco-10. Mac needed an airplane in which he could train students, and haul passengers. Nothing is known about the fate of the Heath or where, when, or how Mac obtained the Waco-10, but he did base it in the field behind the CCC camp.

WACO –10 biplane (courtesy Experimental Aircraft Association Library)

PART IV

DAVID ARRIVES

At approximately 6:00 A.M., March 24, 1937 inside the duplex apartment house where the Heath had been rebuilt located somewhere in Linden, Alabama Jewel McKenzie, nee Jewel Currence of Cuba, Alabama, the wife of George Dewey McKenzie from Laurel, Mississippi gave birth to a son.

It has jokingly been said that they had not chosen a name for the youngster and the cook/housekeeper, upon seeing him said, "My, my, he's so big you ought to name him *Golasha* after that giant in the Bible that David killed." Jewel had a brother who was a career soldier in the U.S. Army named Edwin Elmo Currence whose middle name was chosen as a middle name and so, David Elmo McKenzie joined the inhabitants of this planet, Earth.

During early 1937 a pilot who was stationed at Maxwell Field in Montgomery would occasionally fly an Army Air Corps. Plane into the field behind the C.C.C. camp, rent the Waco and fly on to Whitfield to visit his wife's relatives. There were no fields at Whitfield that were suitable for the Army plane to land in so, he'd stop in Linden and rent Mac's Waco 10. He'd spend the weekend in Whitfield and on Sunday afternoon return to Linden, settle the rental bill and then depart in the military airplane. On the last one of the trips he told Mac that he was going to China and invited Mac to join the group and accompany them. Mac rejected the invitation and later learned that the pilot was a member of the Air Corps. Precision flight team called, "The Skylarks". They had replaced the original team named, "Three Men on a Flying Trapeze" that had disbanded when the leader, Captain Clare Chenault had left the U.S. Army Air Corps. Those teams had been the forerunners of the current team called, the, "Thunderbirds".

Mac had met Captain Chenault in 1934 when he would visit Key Field during the Key brothers' first attempt to set an endurance record that they succeeded in accomplishing in 1935 when they remained airborne for 628 continuous hours. Captain Chenault had obtained and provided the new gyroscopic artificial attitude indicator that the Key's had installed in the Curtiss Robin for use during the endurance flight. Of course, Captain Chenault and the other pilots who resigned from the three branches of the armed forces in the United States and went to China became the American Volunteer Group more commonly known as the, "Flying Tigers".

Mac explained later that the reason for his refusal to attempt joining the group was due to growing family obligations.

To no one's surprise Mac found that the aviation business in Linden during the depression was not lucrative. He interviewed several people in Linden who had expressed interest in aviation and was told that they were intimidated by the size of the Waco. If the truth were admitted it was probably because they were afraid to fly or too poor so, he chose to attempt to lower the overhead by trading the Waco-10 to Dr. Wren Allen and "Stumpy" Armistad in Montgomery for a nearly new J-2 Taylor or Piper Cub.

It is difficult to determine which airplane it was because the Cub was designed by C. G. Taylor and his brother. Then William T. Piper bought them out and started production of the same airplane in Bradford, Pennsylvania calling it the Piper J-2 Cub. It had to be the J-2 though because it had the removable side curtains instead of the plexiglass windows. Mac later stated, "Before I got home from Montgomery with the Cub I was wishing I had my Waco back".

The flying did not improve with the smaller airplane. Mac speculated that the potential students had never intended to fly and that their libidos prevented their admissions of the fact. He attempted to "barnstorm" the Cub that led to a weekend in 1937 or '38 when he took it to Selma in an attempt to sell sight-seeing rides over the flooded Alabama River.

Upon returning to Linden that Sunday afternoon he found Jewel and his son at the field. Since it was not yet sundown he suggested that Jewel take a ride. She refused the offer pointing out that she had the boy with her and Mac rejected her excuse saying, "He's little. Hold him in your lap."

Mac stated later that Jewel addressed the youngster saying, "I'll bet this is one time you'll sit down and stay quiet.". The youngster didn't. Mac claimed that the boy stood in his mother's lap and gripped the window edge during the takeoff, the flight, and the landing. He was completely quiet until the airplane came to a stop and then the youngster protested the termination of the ride. Mac said he had to open the throttle and taxi to silence the boy's protest and each time he closed the throttle the boy would resume the protest.

PART V

A FINAL MOVE

During the summer and fall of 1931 the little town named Evergreen in South Alabama obtained an airport through no effort of its own. Observation and examination of its history would inspire anyone to question the wisdom of putting an airport there because no need or desire has ever existed in the community that justified the establishment of the facility. It was built by the U.S. Department of Commerce because the location is approximately halfway between Mobile and Montgomery and an airway beacon light was placed on a tower about five and eight tenths miles west of the town on the south side of U.S. Highway 84 which was a gravel road that ran from Andalusia east of Evergreen and continued west until it joined the paved road about eight miles east of Evergreen and shared the route with U.S. Highway 31 westward, through Evergreen to the Fairview intersection where it diverged to the right, became gravel again and continued to Grove Hill, a small town, that lies sixty miles to the west. Highway 31 was the only paved road that ran through Conecuh County of which Evergreen is the Seat.

The Civil Aeronautics Authority (CAA) had built a communications and weather observation station at the south side of the beacon tower and staffed the facility twenty-four hours per day for three hundred sixty five days per year. The beacon tower was on the northeast corner of the little grass airport that was approximately two thousand five hundred feet square. The southern boundary was about where the intersection of the taxiways is today and the western boundary has never been moved.

Robert Fowler was undoubtedly the first pilot to bring an airplane to Conecuh County when he passed through on January 11, 1912 flying a Wright Model B which he landed two miles east of town and north of the L&N railroad tracks. He serviced his airplane there from the train that was following him on the first successful transcontinental flight from west to east. It was the second successful transcontinental flight.

In February, 1939 Monroe Sanders had told Mac that there were seven sawmills in a town called Evergreen and there was no one in the entire county that could weld. There was also an airport located there and it had no operator or manager. Mac, being open to opportunity and seeking his fortune moved with the assistance of R. V. McLendon of Brooklyn, Alabama from Linden to Evergreen. The small familie's first residence in the little town was in an apartment in the home of Morgan Long on Magnolia Avenue.

He purchased property from the L. D. King Lumber Co. on Belleville Street south of the railroad crossing and on the east side of the street. There he constructed a metal building in which he opened Mac's Repair Shop that was Evergreen's first welding shop intended to serve the general public.

It wasn't long before he had an arrangement with the Taylorcraft Company of Alliance, Ohio that was producing a two passenger, side-by-side seating, hi-wing monoplane. He sold one to Dr. Carter of Repton, Alabama and another one to Dr. Holly of Brewton, Alabama. His aviation adventures during that period were to pick-up new Taylorcraft airplanes in Alliance and deliver them to buyers in the southeastern states.

TAYLORCRAFT BC-12 (courtesy Experimental Aircraft Association Library)

During those operations he did have one forced landing when the Talorcraft he had sold to Dr. Holly in Brewton, Alabama suffered a catastrophic failure of its Lycoming engine. He stated later that he was more concerned about ripping open the incision from an appendectomy that had been very recently performed on him by Dr. Carter if he mishandled the forced landing.

Neither the airplane was damaged nor was Mac injured in the forced landing and later Mac removed the engine, overhauled it, re-installed it, and flew the airplane out of the field he had put it in. Afterwards Mac would recall how the local authorities attempted to prevent his flying the airplane out of the field. He had a very strong animosity towards the, "self appointed" experts who were not mechanics or pilots.

The thought of a forced landing seemed to be always on his mind. He never exhibited any fear of such an event. It was simply the resolution to never be caught by surprise or in a position where it would be impossible to make one if it became necessary. He often explained that his first instructor gave him a simulated forced landing before he had completed two hours of dual instruction and said, "You'll have a real forced landing before you solo.". As stated earlier, he soloed in six hours and he did indeed, have a genuine engine failure prior to solo.

The leaders of the United States knew that World War II was going to involve the United States and they had begun preparations for it as early as mid-nineteen forty. Mac had given up the general, small town welding shop called Mac's Repair Shop and had converted it to a government supported aircraft welding school that was very successful and produced welders that were employed at Brookley Field, Mobile, Eglin Field, Florida, Keesler Field, Biloxi, Maxwell and Gunter Fields, Montgomery, and some went to the shipyards in Mobile and Gulfport.

Those welding students rented rooms from home owners in Evergreen and came from South Alabama, Mississippi, and Florida. Mac was always very proud of the success that the female students exhibited. Unfortunately nothing was heard from them after they were placed in jobs at the various military installations.

Naturally, there were some, "motorsickles" in the family. While operating the welding school Mac would give David a ride on one after lunch everyday and then return to the school. It was during those rides that Mac established a theory about animal's ability to communicate with each other such as dogs to dogs, cats to cats, cows to cows, etc.

Dogs in Evergreen did not like, "motorsickles". They probably had never seen one before "Mac" McKenzie arrived in town. Every time he went for a ride every dog in town that he passed would chase him and by doing so, endanger him. Mac went to the, "City Drug Store" and purchased a squeegee bulb liquid dispenser and a can of, "Miller Hi-Life", (that's not beer) loaded the squeegee bulb with, "Hi-Life" and went for a ride.

Sure enough, a dog appeared running alongside the, "motorsickle", barking, and attempting to bite Mac's leg. Mac set the squeegee above the dog's tail and proceeded to spray the animal. He had decided that the dog would probably stop immediately when the fluid contacted him so, in order to strike the dog with enough to be effective he thought that if he started at the tail then when the dog stopped he'd cover it from tail to nose. He did. The dog stopped, looked up at him, then ran to the side of the street while howling and when it reached the grass beside the street rolled onto its back and slid on its back attempting to remove the offending fluid. The dog repeated the procedure several times while continuing the howling. Mac said he stopped the bike and could hear every dog in town howling.

He then rode around town and was not chased by another dog. He swore that animals must be capable of communicating with each other, at least within the same breeds. Whether or not animals are capable of communication is still being studied by some academics, but Mac wasn't troubled with any more harassment by the dogs.

PART VI

JAMES LEONARD MCKENZIE

"Jimmy" made his appearance on February 19, 1941 in Dr. Holley's hospital on Douglass Avenue in Brewton, Alabama. He was named after an uncle on his father's side of the family. Jewel has said, "While I was in the hospital recovering from the delivery of Jimmy Mac came in and said, "Honey, I bought another airplane.".

In later years Jewel exhibited considerable bitterness and a lack of appreciation for Mac's selection of that occasion to advise her of his decision to purchase another, "damned airplane" with money they had been saving for the purchase of a house. All the facts about that episode have never been released nor will they ever be.

Jewel said, "He then came home in the middle of one afternoon and announced that he had sold the shop building and was closing the welding school." He then accepted employment as a flight instructor in a Civilian Pilot Training (CPT) school at Lovell Field in Chattanooga, Tennessee.

The flight instructor position in Chattanooga must have provided a respectable income as evidenced by the purchase of a new suit of furniture for another room in the house every time Mac came home. He was allowed to instruct eighty hours flying time per month in the CPT school which he normally completed in two and a half to three weeks and then he'd return to Evergreen and remain until the beginning of the next month.

Jewel bitterly recalled periods when she was running low on groceries and merchants in Evergreen wouldn't allow her to purchase on credit. To this day it is extremely difficult to understand why or how such a problem could have existed and yet life continued with the family reasonably happy. David remembers the women in the neighborhood gathering in the kitchen every morning to share their stories and camaraderie and the cook and housekeeper, Lela Belle Curry who still occasionally visited the family in their home fifteen years later. Although she seemed happy to all the friends and neighbors it was undoubtedly a lonely time for Jewel.

On one of his visits home Jewel and Mac had purchased a record player-radio combination at the local Western-Auto store and while Mac was in Chattanooga Jewel let the teenagers in the neighborhood purchase and store seventy-eight R.P.M. records in the cabinet of the phonograph-radio console. They'd stop at the McKenzie house on their way home after school and dance in the living room until they felt the need to make their appearances at their own homes.

Mac was in Chattanooga on December 7, 1941. David still remembers the living room and front porch being filled with neighbors listening to the radio console and him asking Jewel, "What's goin' on?"

She answered, "The Japs bombed Pearl Harbor."

"What's a Pearl Harbor?"

"It's a Naval Base.

He pulled up his shirt and looked down at his navel while wondering, "What's the big deal?"

He still remembers the same type of afternoon when Jewel's answer to his question was, "Carole Lombard's been killed. She was a movie star."

During one of Mac's absences from Evergreen his son, David had a fight with the youngster residing next door. It was one of the few fights that David ever won.

After the fight George Dean ran home crying and was threatening David by advising that he was going to tell his mother who would come out and "whup" him. That threat caused David to seek a place to hide from Mrs. Dean that would also provide some diversion and entertainment until either lunchtime arrived or time had eliminated the eminent threat of Mrs. Dean's extraction of vengeance.

Mac had three, "motorsickles" in the wooden garage that was normally used as a wood shed in the back yard at that house, one of which he had purchased from the State of Alabama Highway Department. It was a beautiful black and white Harley Davidson police bike that the nearly five-year old David loved to have rides on. He chose the little garage as a hiding place and soon climbed on his favorite bike.

The previous occupant of the house on Belleville Street had left a supply of firewood on the dirt floor of the little garage and the doors had fallen onto the woodpile. Mac rearranged the wood and the doors so that he could ride the, "motorsickle" onto the sloping doors, leave it in gear and parked. Then upon return and removing a bike from the garage he only needed to release the clutch and the bike would roll out by the force of gravity instead of having to exert human effort for the retrieval. Mac was a great conservationist of effort.

David mounted the bike and proceeded to play with the controls and managed to release the clutch. The bike rolled out of the garage rearward, stopped, and David proceeded to get off on the right side pulling the bike over on him. The front "crash" bar was on his right leg and the rear one was on his left foot.

"What a predicament!" he thought. "The neighbor kid's mom is gonna get me. My own mom is gonna' get me. And when dad gets home, whenever that may be, he's gonna' get me." His right leg was beginning to be very painful and he couldn't get out from under the "motorsickle". There was nothing to do but scream. He did and fortunately it was nearly noon. When his mom, Jewel responded to his screams, couldn't lift the bike, got Mrs. Dean to help, and the two of them couldn't lift the bike, they all waited until Mr. Dean came home for lunch and the Deans lifted the bike while Jewel dragged David from beneath it.

David was then taken into the house, placed in bed and, "Dr. Rob" was called. Dr. Robert Stallworth diagnosed the injuries as not being serious and advised Jewel to simply let the boy stay in bed until he decided to get up, out, and about.

Early that evening Mrs. Dean and her son George came to visit and while sitting beside the bed Mrs. Dean said, "I'm sure sorry it happened.", to which George Dean commented, "I'm not. I'm glad he got it." He may have wanted to say more but his mother grasped his ear and propelled him towards the door.

David soon heard the unmistakable sounds of a, "whuppin' goin' on." It was the, "high point" of the day to him.

Late in 1942 the management of the flying school in Chattanooga directed Mac to report to another school in Georgia that they were opening. As the reader has probably decided by this point Mac was a, "free spirit." If he didn't understand and agree with the need for a particular action or decision he wasn't likely to comply with the direction provided. That personality characteristic was perhaps inherited by both of his sons. Mac did leave the home and started the drive to the assignment in Georgia but, as the trip progressed he began to question his decisions.

A few hours after the departure he reappeared at home, called the CPT school management and resigned from the job. The next day he began an effort to obtain a draft deferment from the local Draft Board based upon the facts that he was an accomplished welder and there was not another one in Conecuh County and since the skill was essential to the farmers and loggers in the area he should be awarded one and open the welding shop. He got it.

Jewel didn't want Mac to continue to attempt establishing a career in aviation and she definitely did not wish to relocate to another community. She was ready to establish a home, then and there.

David was very ill again. When he was two years old he had nearly died from pyalitus, a kidney disease. This time he was bedridden for two weeks and then Jewel and Mac decided to take him to a pediatrician in Montgomery. Dr. Harris P. Dawson had listened to a description of the symptoms on the telephone and ordered them to take him directly to Jackson's Hospital where tests were run and on the second day Dr. Dawson entered the room and announced, "He has typhoid fever."

He was hospitalized for typhoid and can still remember the air raid drills, the roar of a lion in the zoo across the street, and the chills every morning. Following the recovery from typhoid he was hospitalized again for a tonsillectomy. Then 1942 came to an end.

Mac made an agreement with Jack Neuman, the owner of a sawmill at the south end of Pecan Street and temporarily established and opened a welding shop in a wooden garage beside the last house on the east side of the street. They soon purchased the house at 217 Pecan Street where they were to live for the next fifteen years. Within a very few months they purchased the empty lot on the south side of the house and erected a forty foot wide by eighty foot long wooden building with no floor, a tar paper covered roof, and asbestos siding made to appear the same as brick walls but failed in the attempt to simulate brick. The window openings were covered with wooden planks and the customer parking lot in front was gravel. It opened at 219 Pecan Street as, Mac's Repair Shop and remained so until the summer of 1956.

No one who ever knew them could ever begin to understand how a couple with such differing ambitions, personalities, and tastes could have ever associated with each other, let alone marry, and start a family. Jewel was reserved except when her temper was aroused which occurred frequently. There is no question that she was very attractive and always carried herself as a lady, albeit a tough one. Although she had mastered profanity at an early age she never exhibited her fluency with that vocabulary except when unbelievably frustrated and angered. It is doubtful that anyone in the county of sixteen thousand people could specify an instance when they had seen, "Miss Jewel" angered or heard her using the ultimate relief for frustration caused by the inability to articulate one's thoughts in a manner that would be impossible to fail to comprehend by any listener known as "cussin". She without a doubt inherited the art form from her father who inherited it from his father who her children had speculated was using the name, "Currence", as an assumed name because he was probably an escapee from, "Devil's Island".

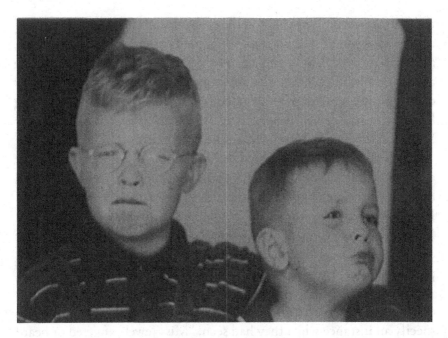

David & Jimmy in 1943 or '44 (author's collection)

PART VII

FORMAL EDUCATION FOR DAVID BEGINS

In the larger towns such as Mobile, Montgomery, Birmingham, and Pensacola a youngster's introduction to a formal education began in Kindergarten at the age of five. In outlying rural communities like Evergreen, Alabama school began at the age of six in the first grade. In September, 1943 David suffered the trauma of having his lifestyle completely revised by entering Mrs. Weathers' first grade class at Evergreen City School with W. Paul (Mr. Mac) McMillan as Principal whom David came to know very well.

There must have been sixty children in the two classrooms, Mrs. Weathers' room and Mrs. Simmons' room that made up the first grade at school in 1943-'44 which on the second day, dropped to fifteen or twenty. The class had an additional informal sub-division into groups called by the children, "town kids and country kids". On the first day all of the students attended and were enrolled, and assigned to sections. On the second day and for the next two or three weeks the "country kids" were excused from school attendance in order to help their parents pick cotton hence, the reduction in class size.

David may have thought that left too many people in his way. Especially, when he was released from class to eat lunch and had to wait in a line extending down the hallway from the lunchroom door which when entered allowed the "diner" to give the attendant seated there his or her name to cross-check against the list of pre-paid diners and if the person's name was not on the list then that person was required to pay fifteen cents cash to the attendant who incidentally, happened to be the lunchroom supervisor, chief, or "boss". After successfully completing that formality the prospective diner was allowed to pick up a tray, eating utensils, a carton of milk and pass through the serving line, after which he or she would exit the lunchroom, sit down at one of the picnic tables in the hallway, eat, then clean the tray, place it in the stack with the other trays, drop the used utensils in the proper can and proceed to the playground. Of course a disproportionate amount of time allocated for the lunch period had already passed into that inconceivable, and only imaginable place where time goes and becomes lost forever by the time a student could accomplish all of those things.

David did not approve of wasting time. Especially that which he considered to be, "his" by standing in line but, being courteous and considerate of the rights of others he would not attempt to break into a line. Obviously, the whole situation could be improved if lines could be moved faster. The day came when he turned around and went to the end of the hallway where he turned around again, focused his eyes and attention on the body of the last child in the offending line, lowered his head, and began his run. On the well-oiled floor of the hallway it must have taken five or six steps to reach terminal velocity.

It may have been the adrenalin rush or the extremely short duration of time that elapsed during the very brief moment of actual impact or even the focus of concentration on the objective but, most likely it was the combination of all those factors that kept David from feeling any pain as he began to feel the satisfaction of the unmoving and unresponsive lunch line yielding to the pressure he had applied. He was unaware of any noise caused by his attempt to shorten the line but, it had shortened.

He was mentally and visually measuring the distance from his prior standing position at the end of the line to the modified position of the line's end when he suddenly became aware of a great amount of pressure on his glutinous maximous muscles. Instantly looking rearward he saw for the first time in his life, a man who was to have a great influence on his activities for years to come. Mr. Mac, the Principle, had just exercised his prerogative of dispensing capital punishment, the objective of which is to encourage behavior modification. Mr. Mac removed David from the lunch line, turned supervision of the youngster over to someone else and instructed that party to hold the boy until everyone else in the school had lunch and then let him in the lunchroom.

By the time David finished eating he only had about ten minutes on the playground, all of which he endured as the center of attention of the other children who from the circle they formed around him said, "David M-u-u-h-k-e-e-nz-e-e-e, A-a-h-h-u-u-u-m, you got a whu-u-u-u-p-i-i-i-i-n.".

It was not a good day. When he arrived at home after school he learned that his mother already knew about the day's events. In future years he came to learn that people in the "public eye", especially those in small towns are unable to do anything without the entire community almost immediately learning of it and judging it and the perpetrator accordingly. When deprived of anything to use in judging a person they will often imagine something to relate. In the case of the crowd control experiment he was reminded for days afterward of the consequences of his attempt to influence society. When the line of children at the door to the lunchroom went crashing through the door they also collided with the supervisor who was checking them in and collecting the fifteen-cent fees. David was told that she had the money tray knocked from her lap, lost all of the money, and several kids probably got into the lunchroom without paying for their lunches. They'd probably never find all of the money that was dropped and the lunchroom might have to close because of the financial disaster.

He really didn't believe it even though he was only six years old and became skeptical of similar claims when he reached voting age.

Mrs. Weathers' husband was the Conecuh County Superintendent of Education and he appeared in the classroom soon after David's crowd control and line shortening experiment and conducted visual acuity tests. David could see no reason for such testing, after all, he obviously had no difficulty in recognizing long lines and selecting targets. There was no question about whether or not he could see but, they said he needed glasses. The trip northward on U.S. Highway 31 from Evergreen to Montgomery in the 1940 Chevrolet two door sedan owned and driven by Mac required about two and a half hours followed by the examination by an ophthalmologist which was followed by a visit to an optician, an exchange of money, and the return trip to Evergreen arriving around sundown. The U.S. Mail carrier delivered the glasses in about two weeks.

The temple pieces running from the lens frame rearward and wrapping around the ears caused painful irritated areas on the back of his ears. The nose pads irritated his nose and he couldn't really detect or perhaps care whether he could see any better but, the glasses did have some benefits. He was told to stay out of fights, not play ball, don't do anything rough, and to, "be good". He didn't object to those recommendations after all, Ward beat him up, Richard beat him up, and Wayne beat him up. The truth is, he seldom won a fight. He had never seen or even heard of playing with a ball before entering school and that looked like a great bore and waste of time. Besides there wasn't much going on. It was just a lot of standing around and waiting for something to happen so, there was no objection to foregoing participation in that foolishness. When his ears "toughened" to the temple pieces and his nose adjusted to the pads wearing glasses wasn't really so bad except for the prejudice displayed by other members of society.

Prejudice is ALWAYS founded in and supported by ignorance and so it was in David's case. He was always being told by other children, "You can't play. You have to wear glasses."

One afternoon when the parents were visiting the classroom to observe their, "little darlings'" progress the question being considered by the students was, "What do you want to do when you grow up?" David's response was, "Fly airplanes."

"NO, NO, NO, you can't do that. You have to wear glasses and airplane pilots can't wear glasses." was the response he remembered. He couldn't recall who said it but, it was loud, authoritative, and forcefully clear. "Now pick something else that you'd like to do."

All eyes were on him. Teachers, parents, students, the whole world was focused on him and he hated being the center of attention in addition to being singled out and exhibited as an example of an overconfident and ridiculously ambitious misfit. In an effort to save himself from the objectionable situation he finally said, "I want to work for Mr. Ford."

The class session proceeded with no further difficulty and David was not called upon again for class participation that day. He had satisfied the demands of everyone by making a life forming decision at the age of six years old which was truly remarkable for a person so young in South Alabama. It is additionally remarkable that he did in fact leave Alabama, joined Ford Motor Company at the age of twenty-eight and retired after thirty nine years and eight months employment by the company.

Mac purchased a, Bruner-Winkle "Bird" three-passenger biplane from someone in Gainesville, Georgia and returned with it to Evergreen one weekend. The next Sunday morning he flew it to the airport at Brewton. The airport was located on the north side of U.S. Highway 31 on the northeast side of town and was also the local golf course. The airport, the hangar with its accompanying lounge and restaurant on the first floor, and dance floor on the second level had been built during the depression by the Works Progress Administration (WPA) and was the type facility that would have been envied by the citizens of any community. It had one grass runway running from southwest to the northeast and a second runway perpendicular to that. The golfers shared the runways with the airplanes and agreed to evacuate a runway whenever they saw or heard an airplane attempting to use it. That agreement appeared to work fine. There were no collisions by airplanes and golfers and no one was ever seriously injured or killed at the site while attempting to play a round of golf or, so far as the author knows flying airplanes either.

On most Saturday afternoons Jewel and Mac would put the boys in the Chevrolet and drive the thirty miles south to the airport where Mac would sell rides and teach students. They repeated the activity every Sunday that weather permitted but, Sundays were all day affairs. David could remember his mother sitting at a card table in the front of the hangar selling rides in the Bird, placing the proceeds in a cigar box, and escorting passengers to and from the airplane each time his dad landed. She usually had both sons with her. Mac claimed he did well at Brewton. He had the only airplane on the airport capable of carrying two passengers at once on the same flight. Couples preferred that and Mac enjoyed the benefit of their preference. Those were GOOD TIMES.

BRUNER-WINKLE "BIRD" with 100 H.P. Kinner engine
(courtesy Experimental Aircraft Association Library)

During that period a resident of Brewton started flying and unfortunately, he suffered an engine failure on takeoff towards the southwest one afternoon, stalled the airplane, and fractured a wrist in the accident. Bill Blacksher had about one hundred sixty hours experience in his career at that time and was to meet Mac and David fifteen years later.

During that time the U.S. Navy acquired ownership of the little airport at Evergreen, purchased property adjoining it from the Ward family, built the two four thousand foot long paved runways with parallel taxiways, the concrete parking ramp, and the connecting taxiway. The two runways had boundary lighting that was flush with the surface that when viewed at night with the rotating beacon sweeping alternating beams of green and white light around the sky was a beautiful sight.

The Middleton family resided on the south side of U.S. Highway 84 approximately two miles east of the airport and had a son, Laula who flew B-25 bombers during World War II. Laula Middleton lost his life in combat in the Mediterranean so, the airport in Evergreen was named, "Middleton Field" in his honor. The U.S. Navy also built the airport on the south side of Brewton to serve as another outlying field for training operations based at Whiting Field, Naval Air Station near Milton, Florida but, World War II ended before the Navy could begin operations at the two airports. Ownership of the airport at Evergreen reverted back to the City of Evergreen and as the City officials allowed it to degenerate kept its name, "Middleton Field".

Mac McKenzie had sold the "Bird" he had operated at Brewton and begun to operate at Evergreen and sometimes from an airport located south of Andalusia, Alabama when the Patten family in Evergreen contacted him about the possibility of teaching the head of the family, Eugene to fly. Little is recalled about Mr. Patten's introduction to aviation except that both families were in Andalusia when Mac took Mr. Patten for a flight and then took Mrs. Patten and their son, John for a flight. The airplane was a two place side by side seating machine that was probably a Taylorcraft although it may have been an Aeronca Chief. John is approximately three years younger than David and has said that he still remembers sitting in the airplane as David explained the function of the controls. John also relates how his mother recalls a story about him admonishing her during their flight saying, "Mom, be quiet. I'm trying to look." That flight probably occurred during '44'or 45.

The Spirit's Journey

PIPER J-3 CUB (courtesy Experimental Aircraft Association Library)

AERONCA 7AC "CHAMPION" (courtesy Experimental Aircraft Association Library)

Some of the airplanes owned by Mac and based at, "Middleton Field" were a J-3 Piper Cub, a 7-AC Aeronca Champ, an L-2 Taylorcraft, and he owned the Bird for a second time during that period and sold it to Ansle Wilson who had married one of Walt Ward's daughters. Walt Ward was part of the family whose property was purchased for the airport expansion.

Some of the people who had contact with, "Middleton Field" during the post war years were Ansle Wilson, Ansle's brother, Maxine Thompson, and Ben Carpenter. Miss Thompson and Mr. Carpenter were taught to fly by Mac in the J-3 Piper Cub.

One day in the summer of '46 while they were all at the airport Jewel approached Mac and said, "Take David for a ride and scare him."

Mac had no difficulty persuading David to get in the front seat of the Cub as Jewel frowned and walked away all of which pleased the youngster. Mac was more careful than usual inspecting the youngster's seating position and the fitting of the safety belt. Of course, he had to hand prop the airplane to start the engine and in David's opinion there never could be any other odor that was more pleasant than that of the exhaust of a smoothly idling small airplane engine on a warm and still summer afternoon being pushed rearward over and around the fuselage by the propeller and picking up the odors of eighty octane aviation gasoline and the nitrate and butyrate dope painted onto the fabric covering of the machine. Mix those odors with those of grass, clean air, and farm fields and the resulting mixture is soothing, intoxicating, spell binding, and absolutely impossible to describe. The impression, the mood, and the wonderful feeling was transposed to one of exhilaration and joy as the airplane started its takeoff run and the moment the tires lifted off the ground it felt like spirits must feel as they drift through time and space with absolutely no restrictions. It was far from the first time David ever flew and certainly not the last time he'd ever fly but, he would never forget the sensation of freedom that he felt every time an airplane lifted off the ground and provided the sensensation that some tether had been broken that was restraining him to an unknown unidentifiable and objectionable force which limited his freedom.

Mac with David in the front seat of the Cub climbed out of the airport traffic pattern with the side door and window open. When they reached sufficient altitude Mac closed the door and window and said, "Hold onto the seat frame so that your arms don't swing about." David followed Mac's instructions and being small he couldn't see over the instrument panel or downwards over the door edge or the lower edge of the left side window, but he was thrilled by the view of the earth sometimes appearing in the windshield, sometimes in the skylight in the roof of the cockpit, sometimes in one or the other side windows, and the physical sensation of varying "G" forces on his body accompanied by changes in visual acuity and hearing as changing positive "G's" had their effect.

Because the Cub didn't have a fuel system that would function while inverted and using good judgment, Mac didn't perform any negative "G" maneuvers. The engine never quit running and never over sped. The forces of acceleration were never very high because Mac was performing the aerobatics in a graceful, sweeping, very pleasing in appearance manner that would keep "positive G" forces on the airplane and its passenger. If anyone on the ground had been observing the flight they would have seen a performance that was careful, safe, pretty, and entertaining.

David loved the ride. Jewel never said anything about it even though her strategy had failed. Neither did Mac. In later years David often wondered how people could fear aerobatics and yet pay for rides on the shaking, dirty apparatus found in country carnivals called rides and claim to enjoy them.

Mac purchased a small piece of property southeast of the runway intersection from Walt Ward and constructed a single "T" hangar of concrete block on the site. Herman Patten, John's older brother assisted Mac in the construction of the roof on that hangar which was blown off in 1950 by a hurricane, which remarkably, did not damage the Luscombe 8-A that Mac had in it at the time. The hangar was never repaired.

There was a time during the early 1950's when the Bird was tied down beside a Meyers OTW biplane at Mac's former hangar site. Ansle Wilson had damaged the Bird in a forced landing when he mismanaged the fuel supply. The wings for the Meyer's OTW had been recovered by Mac for the owner, Alva or Bill Register of Troy, Alabama. Mr. Register may have been incarcerated at the time because of some activities related to the first administration of Governor James Folsom. Mac was never compensated for the work but, information on all of that is very vague.

During 1949, '50, and '51 Mac would occasionally give David some, "dual instruction" in the Luscombe but it was instruction in the, "air work" phase of pilot training only and undoubtedly did not contribute to David's later accomplishments.

That Luscombe was the last airplane that Mac owned because of the pressure and resistance from Jewel. She had developed a strong and sincere hatred and distrust of pilots, airplane owners, and anyone else associated with aviation. She frequently expressed her views and opinions on the subject by saying, "They are full to the neck of tall tales meant to show you how great they are. They're lies, all lies. When one comes through the door he looks you straight in the eye and offers his hand so that you won't notice that his other hand is covering the seat of his britches so that you can't see the hole that's worn through them and know that he doesn't have enough money to buy a new pair. They're all bums."

Luscombe 8-A (Shown with "rag wings". Mac's had the all metal wings) (courtesy Experimental Aircraft Association Library)

Since he was, "laid back" and easy going, Mac yielded to the demands from Jewel. He probably sensed that Jewel's objection to aviation had reached the level where continued efforts and expenditures in the activity endangered the family unity.

Jimmy was not a healthy baby and naturally did not have the demands placed on him that David faced. At least, that's what David thought. While Mac was working in Chattanooga Jewel would put the boys to bed early each evening and David recalled how he could still hear the other youngsters in the neighborhood playing outside while he was being forced to go to bed. His mother blamed the failure of Jimmy to immediately go to sleep on the disturbances created by David. Her solution to that problem was to force David to rock Jimmy's crib until he went to sleep. David didn't care what silenced the little guy so long as he simply became quiet. He hated having to rock that crib and sometimes would lie on his back beneath it, place his feet on the springs, and rhythmically push the springs up, then allow gravity to lower them. It isn't known whether it was the rhythmical movement or fear that quieted the baby and David didn't care, but it created peace in the household.

After they moved into the house on Pecan Street Jimmy became just another kid in the neighborhood and the rest of the family fell into a routine existence. Mac would never voluntarily get out of bed early in the morning. That is prior to 9:00 A.M. and by then the parking lot in front of the shop would be full of potential customers who would hold a meeting, select a delegate to go to the back door of the house, knock, and when Jewel answered, ask, "Miss Jewel, is Mac up yet?"

Jewel's response was usually, "No, he's still in there in the bed. Come in and you can talk to him." There is no record of any customer refusing that invitation and it seemed to become a standard and accepted procedure when doing business at Mac's Repair Shop.

In the southern part of the United States twelve o'clock, noon is "dinner time" in all homes. It's lunchtime in schools, government institutions, and any business with national ties but twelve o'clock is, "dinner time". Even though on most days Mac didn't open the shop until 10:00 o'clock he closed it at 12:00 to go to dinner and if anyone was not already aware of it he would advise them of the dangers of returning to work while the stomach was struggling with its duties and function. The dinner break ran from 12:00 until 2:00 P.M. and of course 5:00 P.M. was closing time for all businesses in Evergreen and that included Mac's Repair Shop regardless of the amount of work that may have been completed that day. In other words approximately five hours was a full day's work in the shop. Even though the house was next door to the shop, with the exception of the morning customer's delegate very few people ever went to the house looking for Mac, especially if the visit was job related.

There were two businesses in Evergreen at that time that enjoyed the comfort of air conditioning. One was the, "Bank of Evergreen" and the other was the, "Pix Theater". Residents of the community locked their doors in the evening, but slept with the windows open and fans running on most summer nights. Everything was open during daylight hours. All of those conditions combined one afternoon to illustrate Jimmy's openness and integrity when a potential customer arrived about 1:30 in the afternoon and found Jimmy who was four or five years old playing with his toy cars in the parking lot for the shop. The customer who was of obvious African ancestry stopped his car beside the youngster and asked, "Hey, boy, is your daddy around?"

Jimmy responded, "Yeah, wait a minute, I'll get him." Whereupon he walked to the bedroom window and shouted, "Daddy, there's a nigger out here, wants something." and returned to his toys in the sand.

The customer then asked, "Boy, what did you call me?" Jimmy asked, "Huh?"

"Did you call me a nigger?" "Well, ain't that what you are?" was Jimmy's honest response? Mac arrived at that and it was the end of the "confrontation" which shows the absurdity of displaying intolerance for socially accepted and commonly used "slang" terms. Jimmy never gave the manner any more thought. He did not intend it as an insult and it was not one. It was simply the only description of a homosapien of African ancestry and he used it in honesty and innocence. Why think of it any other way?

Jewel was, "on David's case" one afternoon in his bedroom about something that he had done but should not have, or perhaps something that he should have done but hadn't. Her audio level was higher than normal and her tone was more stern than required for normal conversation but, David could tell that justice would be delivered this time by verbal means, not physically if he would just keep quiet and be patient when he observed that Jimmy, who was still four or five years old had entered the doorway, stopped, and was standing there dressed in his usual pair of short pants and nothing else, leaning on the door casing with his left leg crossed in front of his right leg, obviously awaiting a break in the tongue lashing that David was enduring.

When Jewel finally had to hesitate to take a breath she turned and saw Jimmy in the doorway. He then verbally offered,

"Sheeeeeit."

David was glad to see him that time, as it redirected Jewel's attention.

It appears that there was never a boring moment in the McKenzie household but, that was certainly not the case. It is said that, "Flying is an almost infinite number of hours of pure boredom punctuated by an occasional moment of total terror." That statement is not true or correct. It is always a joy and a very pleasant experience to fly. It is not boring at all. It is very easy to find something that should be done while flying that'll keep you, the pilot busy hence, drive boredom away. Those moments of terror DO NOT occur occasionally. Instead, they rarely occur and there are pilots who have never experienced one in an airplane.

That's not unlike family life and for Jimmy and David it was obviously a given fact that boredom can only exist if the sufferer allows it. Neither of the boys were, trouble makers or mischievous but, they were creative in their assaults on boredom which, like in all families, left memories of grief and memories of delight, humor, joy, and satisfaction.

When Mac opened his welding shop in 1939 he had set the rate for the ordinary small welding job at seventy five cents and when he gave the business to Jimmy in 1968 he was still charging the same rate. Consider the fact that when a customer brought a job into the shop the very first formality to be dealt with was to respond to the customer's enquiries about the condition of, "Miss Jewel", the boys, and flying. Then they had to exchange the latest jokes. This was done as a courtesy similar to the custom in some Asian cultures of bowing to each other upon meeting for the first time in a day. After all of the greetings and salutations the small repair job could require any time from five minutes to a half hour depending on many variables such as how well Mac and the customer knew each other, how complex the repair job, and telling some more jokes. The truth is there is no way that Mac could have earned very much money in the manner he operated the business even when it grew to a staff of three employees and they opened it at 8:00 o'clock. The overhead costs probably kept Mac's personal income as low as ever.

Jewel never commented on the activity or lack of the same in the shop. She stayed out of it and it was a rare occasion when she ever entered the building. During the hot summer months when there came the mid-afternoon thunderstorm or rain shower David and Jimmy could go in the house or often to the shop where Mac had set aside an area located high and to the west side of the personnel door where they stored the tools they were allowed to use. There was a pry bar, a hammer, and a saw that they used. No one else was allowed to use theirs and they couldn't use any belonging to someone else.

Jimmy and David took the tools into the area behind the shop that was overgrown with weeds and started building a fort from wood and junk that they found on the property but, David soon lost interest in the project and quit working on it. Jimmy completed it to his satisfaction. David thought it appeared atrocious. It had two, stories compressed into an overall dizzying height of perhaps, five feet. The upper story was an observation deck with its floor about three feet above the ground and walls on all four sides two feet high for its occupant/defender to duck behind in the event of attack by other kids in the neighborhood. The entrance was through a ditch surrounded by junk, debris, and miscellaneous refuse that a seven or eight year old child could haul, carry, or drag to the site. Ingress was achieved by crossing the debris around the ditch, lying on the stomach in the ditch, and sliding along the bottom and under the wall like a reptile.

Once inside, the occupant could maneuver on his or her hands and knees and if that wasn't satisfactory the option of, climbing to the observation deck was available if the potential for having to sustain an assault attempt by neighboring youngsters from such an exposed position was acceptable.

The appearance of the fort could be described as being, at best horrible. Jimmy liked it. Fortunately, it was constructed in the backyard of the shop and Jewel hadn't noticed the development or if she had she didn't object to it. Mac didn't care whether it was in the rear, side, or even the front so long as the, "bureaurats" and their playmates, the local politicians didn't reappraise the property, require a building permit, or in some other manner put their, "paws" in his pockets. Jewel's opinion changed.

Jewel was looking for Jimmy and he wasn't as easily found as she wished or he was avoiding her deliberately or he may have known why she was looking for him. No one knows but, her search had become noticeable to the customers at the shop. When she finally found him Jimmy fled to the sanctuary of the fort, crossed the debris barrier, dropped into the entry ditch, and scurried underneath the wall to a secure position inside the structure.

Jimmy crouched inside his sanctuary peering out of a portal and neither said nor did any thing as his mother proceeded to try and persuade him to come out. Jewel's pleas did inspire response from Mac's customers in the shop. They came out to see what was happening, saw it, were amused, and stayed out to watch. Jewel was also amused, therefore her tone was undoubtedly not as threatening as a literal interpretation of the words she uttered would have led anyone to believe. Jimmy remained in his fort.

Jewel demanded that he come out. He didn't, even after Jewel threatened to, "go get the belt" if he didn't come out. He didn't move. Observers could see his eyes and nose behind the portal. She told him, "She'd come in there and get him and he'd be sorry if he didn't come out." He didn't move. The observers from Mac's Repair Shop started laughing which probably encouraged Jimmy to continue his resistance.

Jewel was amused, too, so she turned away and started towards the back door of the house while trying to prevent Jimmy from seeing her laughing because that would have completely destroyed the credibility of her threatening demeanor without which she believed she'd have little if any control over the boys.

Dinnertime came and Jimmy appeared at the table. Nothing was said about the fort or Jimmy's failure to come out and face Jewel's discipline, or run her errand, or whatever she wanted. David wondered how that little runt could get away with such shenanigans when he would be the objective of a verbal and physical sentence if he tried it.

There were moments of grief in the family that were caused by both boys but, fortunately, they were always separate events, never simultaneously. David had contracted pyalitis, an infection that caused bleeding in the left kidney when he was two years old, typhoid when he was five that almost cost his life but he grew healthy and well.

VIII

JIMMY STARTS SCHOOL

It became obvious when he started the first grade in school that something was wrong. Jimmy exhibited a lot of difficulty learning to read and count which didn't seem to concern him. His parents and the teacher simply could not understand his lack of progress and Jewel started working with him every afternoon on homework trying to bring him, "up to speed". She would read the assignment to him, have him try to read it to her, allow him to work on it alone for a while, then return and begin the entire process again. It was a daily ritual for five days per week and at least two hours long each day for the extent of his grade school career.

When he reached the end of his first year in school Jimmy, fell ill and the physicians in Evergreen recommended hospitalization in Greenville, Alabama at the Stabler clinic or Hospital where he underwent observation and testing that resulted in the diagnosis of leukemia. It was then recommended that he be examined in Birmingham at The University of Alabama, Birmingham Hospital. Of course, Jewel and Mac drove him to Birmingham as quickly as arrangements and an appointment could be made.

Jewel was so concerned, worried, and depressed with what she was sure was the impending loss that she was unusually quiet and didn't express any thoughts about a course of action to take. Mac was simply quiet. He was doing all that a man could do.

David was too immature to appreciate the real gravity of the situation and thought, "Well, he's not dead yet. Let's see what happens in Birmingham. There's nothing else to do anyhow." David was left at home with friends of the family while Jewel, Mac, and Jimmy went to Birmingham where a bone marrow test was made and the physician announced, "It's not leukemia. It's mononucleosis, *the kissing disease.*"

David could tell by the facial expressions that were visible as the car came down the driveway that everything was going to be fine. After being told of the diagnosis and prognosis he wondered, "Who's been kissin' him?" Jimmy was restricted to bed for the rest of the summer of 1947 and adult visitors would always give him a few coins that he'd add to the collection in his bank, smiling as he did so. That was probably the beginning of one his lifelong passions, saving money. The other was completely inexplicable. He began to collect books and photos for youngsters of his age related and devoted to what he called, "old cars". Then he returned to school in the fall where the studying habits and procedures that were developed the prior year were reinstated.

Marcus Blair was a well driller who displayed a free spirit. He never worked for anyone else and while working for himself had his own schedule, which meant that he worked when he was in the mood, when he felt like it, and when he wanted money. Marcus married Ada McCreary, a nurse, who was a beautiful woman that always seemed to understand and accept the realities in life without displaying emotion. Residents of Conecuh County pronounced her name, "Addie" as did her family.

Ada and Marcus Blair had a son, Wayne, and eventually two daughters. Wayne was approximately the same age as Jimmy and during the summer months when school was not in session Wayne would be with his dad, Marcus when a visit to Mac's was on the agenda. That was often.

Marcus's standard greeting when he entered Mac's shop was, "It's too cold or too hot, or too wet, or too dry, or too dark, or too bright, or too nice a day to work today isn't it, Mac?" Select a reason. Whatever reason selected would describe the day.

Mac never disagreed and would usually join Marcus beside the stove or in front of the fan depending on which one provided the least discomfort unless he was currently serving a customer. They spent a lot of time together and became very close friends.

That work schedule also allowed Wayne a lot of time to spend with the kids on Pecan Street, especially, Jimmy McKenzie. The two of them had heard Mac telling stories about flying and decided that they had to try it but, they didn't have a, "flying machine".

Desire is one of the greatest motivators in the psyche of homosapiens. Desire has caused bar room brawls, street fights, personal conflicts, and even total wars between nations. Perhaps desire was one of the motivators that caused man to communicate, fornicate, and propagate, and desire to have them may have caused the evolution of clothing which is the first clue one sees that distinguishes man from the other animals. Desire probably inspired the construction of housing, and not the least, the development of machines for transportation. It was the desire to fly that inspired Jimmy and Wayne to design and build their, "flying machine".

The two boys raided Mac's scrap pile located behind the shop and retrieved a discarded pedal car and the frame made of angle iron from a home made lawn mower that still had two wheels on it, "sweet talked" Mac's employee into welding the lawn mower frame under the rear of the pedal car body and they had the major component, the fuselage of their flying machine complete. They then attached strips of lumber to the top of the lawn mower frame extending outward from each side of the pedal car body and tacked pieces of cloth torn from an old bed sheet onto the wooden strips.

The, "Jimmy-Wayne" flying machine was ready for its first test flight. Since Jimmy had the benefit of the knowledge and experience gained by riding with his father in airplanes it was agreed that he would be the, "test pilot". He was in the, "pilot's seat" as Wayne pushed the contraption across the parking lot in front of Mac's Repair Shop intending to cross the street and go to the top of the slope in the driveway to Mr. Strong's factory. They thought the assistance of gravity contributed by going down that slope would assist in acquiring flying speed.

Henry was a farmer owning the property that he resided on at the south end of Pecan Street that included the farm fields, the barn, and his house. His family included a wife and three children, all of whom were in a wagon being pulled down the street by a mule and approaching Mac's shop as the boys were departing for their, "test flight".

Until the mid 1960's every street except three in the little town of Evergreen was "integrated" with houses owned and occupied by

Caucasians located consecutively along both sides of the street and suddenly the next home passed by a visitor would be owned and occupied by African Americans as was every home passed after that. The term used to identify the areas occupied by African Americans was, "the Negro quarter". Pecan Street had one African American family residing on it which was headed by Henry - - - - -.

The boys stopped at the end of the driveway entering Mac's Repair Shop to allow the mule and wagon to pass. The mule had never seen such a vehicle therefore, it must not have been comfortable being near the strange beast that was apparently devouring a child. It stopped, raised its head and opened its eyes wider than would be thought possible, laid its ears back, and preceded to back up. Henry - - - - - - started issuing commands in a loud and authoritative tone of voice as the mule continued to retreat rearward which rolled the right rear wheel of the wagon up the trunk of an oak tree raising the rear corner of the wagon dangerously high. Henry's wife and children began to panic and scream.

Upon hearing the disturbance Mac and Marcus stood up and ran out the front door of the shop to investigate. With the windows and doors to the house open Jewel easily heard the cries of panic and fear as she worked in the kitchen, dropped what she was doing, and ran to the front porch where she could see the entire episode developing including Mac and Marcus who by that time were helpless with laughter. Jimmy was sitting in and Wayne was patiently standing beside the "flying machine" and awaiting the next response from the mule before continuing their journey to the site for the test flight.

Jewel began to simultaneously berate and direct Mac and Marcus saying, "Don't just stand there like a pair of buffoons. Go get those babies away from that street. That mule's gonna run over 'em and kill 'em. This is not funny. MOVE!!! If that mule runs away, wrecks that wagon, and kills all those little pickaninnies we never will get it paid for. Get those babies and that mess back over here." Actually, she said a great deal more than that but, it's impossible to recall all of her comments.

Mac was trying to point out that the mule was not going to approach the kids as long as they remained with their vehicle which they were not going to abandon but, Jewel couldn't hear or wouldn't listen to him because of the noise she was making. Her verbal assault eventually over powered the humor the two men found in the situation and they yielded to her demand that they "get those babies and that mess."

After the two men got, "those babies and that mess" out of the way the mule trotted down the street in the direction it was originally going with the three children in the wagon staring back at the site of the commotion while their mother and father were taking frequent looks rearward with shocked expressions on their countenances and the mule looking to its left and rearward, probably to be sure that the strange beast was not following.

The, flying machine was "taxi tested" later and some attempts were made to fly it. Those attempts were never successful of course and the device probably sustained a structural failure. It met its destiny in the scrap pile behind Mac's shop from whence it had risen like the, "Phoenix", thanks to the efforts on its behalf by its believers and creators, Jimmy and Wayne.

David was simply another kid in the neighborhood doing the same things that other youngsters did. There were three nights per week when Jewel and Mac took both boys with them to the, "picture show" at the Pix Theater on West Front Street which has left its own set of memories.

The Pix Theater dominated the building in which it shared occupancy with the People's Shoe Repair and the City Café with the theater in the center, the shoe shop on the west side, and café on the east side with an alley beside the café through which the "minority" citizens were allowed access by going down the alley beside the café to the theater's side door, purchasing their tickets, entering, purchasing their concessions which had been carried from the stand in the main lobby to the side door used by them, and climbing the stairs into the balcony. Of course the Caucasian "majority" used the ticket office on the sidewalk at the front of the building, the concession stand in the main lobby, and the auditorium on the ground level that had a cement floor sloping from high in the rear to low at the stage and viewing screen. That design feature, the sloping floor, led to some audience participation during some years when in the months of April and May as many as six rows of seats would be flooded under water and the patrons would have to make some effort to sympathize with the image on the screen of the cowboy who was stumbling across the desert with no horse, no water, and very little strength remaining under the burning sun to a chorus of croaking frogs.

The Pix Theater was the predominant provider of entertainment for the family since it was enclosed and, on most winter evenings warm. On Tuesday and Thursday evenings Jewel found something that she believed needed her attention while Mac proceeded to read the latest issue of "True" magazine or one of the "True Detective" magazines that he was always bringing home. The boys would play dominoes or checkers with Mac but, never with each other. If they attempted to play any game with each other and without adult supervision there would have been a disagreement, but they never fought physically. They were both clever enough to harass each other verbally and antagonize each other to the point of desiring a fight, but the antagonist would always withdraw prior to blows being exchanged probably with some feeling or sense of satisfaction.

"Bedtime" was nine o'clock unless the family was attending a movie. Even on those evenings the boys played a game invented by their parents to encourage them to go to bed. The first one to get in the bed was allowed and perhaps even encouraged to call the other a "rotten egg" without becoming the victim of parental guidance or discipline. David being the older of the two didn't really participate in the game. He just tolerated the noise from Jimmy. It was the only insult that Jewel and Mac allowed the youngsters to hurl at each other, but be assured that both of them were creative enough to invent additional ones that their parents most likely never heard.

The boys shared the bed in the room at the rear of the house that had a desk, a chair, a chest of drawers, and a couch in it. There was also a night stand on which an antique radio was placed. The room had a closet and enough space was available for Jimmy to set up his toys on the floor at the foot of the bed. The heat was provided by a free standing butane gas heater on the hearth in front of the fireplace that had been closed with a piece of plywood over the opening.

The radio was powered by two dry cell batteries and had two tubes in it. Tuning was by one lever on the top that was simply swung left or tight to select a station and instead of speakers, it had headsets that the boys placed on their heads after going to bed. They'd go to sleep with the radio playing and their mother would remove the headsets and turn the radio off every evening after they fell asleep. The radio was made by the Westinghouse Company in 1924 and had been retrieved from their grandparents home in Laurel, Mississippi during one of the week-end visits and it was believed to be either the first or second radio to be used in a home in Laurel.

One of Jewel's observations about visiting relatives was, "You shouldn't live closer than three to four hours driving distance to your closest relatives and shouldn't see them more often than three times per year and then for no longer that one and a half days. Then all of you will get along fine."

PART IX

DAVID MOVES UP

Probably all of the schoolteachers that David had in school knew of his love of airplanes, but none of them encouraged him to pursue a career in aviation. After all, he wore glasses. Of course it is doubtful that very many teachers in elementary schools consider lifetime vocations. His school life was so nearly normal and routine that it's difficult for him to recall any significant event or occurrence in the second through sixth grades.

The Toliver House
(author's collection)

His second grade teacher, Mrs. Donovan lived in a house on Belleville Street that was called, "The Toliver House". It had four huge columns on the front and was called, "an Ante Bellum home" in architectural language. Every afternoon while walking home from school David passed by the Toliver House and he liked it so much that he vowed to someday reside in such a house.

One morning Mrs. Donovan dispatched him to take the "lunchroom report" to Mr. Mac's office. Upon arrival and presenting the report to the Principal he was asked, "Do you go to church?"

He had no response and while he stood there trying to think of something to say, Mr. Mac said, "Tell your mother to bring you to church Sunday."

After arriving at home that afternoon he rounded up his courage and told Jewel of his encounter with the Principal that morning which led to the McKenzie family appearing at the Sunday morning services of the Evergreen Presbyterian Church.

The Presbyterians made up a small congregation of approximately thirty adults who were allowed to use the St. Mary's Episcopal Church building located on the northwest side of the Evergreen City School building on McMillan Street for Sunday School every Sunday at 10:00 o'clock and on one Sunday per month they could also hold a worship service at 11:00 o'clock conducted by a minister who drove to Evergreen from Brewton to conduct that service. At that time David was aware that the other youngsters in the school that he attended went to Sunday School and Church services every week and seemed to enjoy doing so but, there were few children in the church he started attending once per month and it just didn't seem to be much fun. He used what he believed was good judgment and suppressed any complaints he may have had and went through the motions of going to church once per month. Actually he didn't have any choice.

While he was attending the third grade in the school system from September, 1945 through May, 1946 he and a classmate, Jan Hendrix began a friendship that was to last for the rest of their lives even though no one had any idea what they shared of mutual interest or in common with each other and Jan also attended the Presbyterian services. That was probably because they were held less than a block from his home. When Jan went to Sunday school and Church his younger brother,

Kirven and their dog, Chubby accompanied him. Chubby would follow the boys into the church and lie down beneath the pew they were seated on and no one, absolutely no one made any attempt to evict the dog from the services. That was probably because the dog's deportment was as acceptable as that of any of the churchgoers, worshipers, and attendees. It didn't snore.

In time Jan and David joined the Presbyterian Church in St. Mary's Episcopal Church's building in a ceremony in which David was simultaneously Baptized by Dr. James Gailey. Mac was made a Deacon and the congregation announced its intention to build a church. (That's always going on.)

The leaders of the little congregation counted their pennies regularly, sought and acquired additional members, and pleaded for financial pledges, all of which activities seem to be occurring in churches regularly, anyhow. Then they bought a lot and Mac's father, John D. McKenzie who resided in Laurel, Mississippi and was an architect, among other things, made an initial drawing depicting a proposal for the building's style which was ultimately adopted by the Church Officers and Minister. Just as is the case in most charity, social, and church related fund raising drives, and you can't have one without the other, advantage was taken of personal egos.

The contributors who offered the higher amount of pledges were honored with a metal plate that has their name etched into it attached to one of the window frames in the building and families or persons who contributed a lesser amount have a similar tag attached to the end of a pew. The McKenzie family name is on a metal plate attached to the end of a pew. Mac also constructed some wrought iron railings that were on the building that may have been produced for the cost of the material. Of course, the construction of the Evergreen Presbyterian Church extended over several years. Egos are costly things to have and maintain in exchange for very little physical return.

While in the third grade David started learning the multiplication tables being taught by Mrs. Petrey who became ill necessitating the substitution of Mrs. Gaston who gave him the first lesson he could later recall in the futility of resisting social traditions.

February 14 is two months and one day prior to the mandatory filing of United States Income Tax Return forms, the anniversary of the elimination of seven criminals by their contemporaries in Chicago, and the socially accepted, traditional celebration of a commercially endorsed holiday that is beneficial to jewelry, flower, greeting card, candy, and confectionary shop owners called, "St. Valentine's Day".

Traditionally on the afternoon of February 14 at Evergreen City School every student in each class was expected to provide a collection of Valentine cards addressed to every other student in the class and one to the teacher. There were sixty four students in the class and when the teacher is added and the contributor is deleted from the count in the class that David was in there still remained sixty four cards to be addressed by hand, a return address to be completed on sixty four cards, and the word, "from" to be written by hand between the name of the addressee and the name of the addressor on sixty four cards. The initial acquisition of the cards was performed by visiting V. J. Elmore's 5c, 10c, 15c, & $1.00 Store on West Front Street east of the alley beside the City Café and purchasing the required supply which was not easily done be a youngster whose pride made it difficult to ask his parents for the funding required to execute the purchase.

In spite of the necessity of conquering his pride David had acquired the supply of cards he needed but, he postponed the chore of addressing them until the evening of February 13, 1946, the day before another day that he later thought should live in infamy.

He began to address the Valentine cards and decided that the production rate was slow so, he established a streamlined production system. He went through the remainder of the mountain of cards writing the first name of the addressee on each, then writing a last name on those cards addressed to recipients having identical first names, then adding the word, "from" on the cards by going through the mountain of cards again, and then starting to write his own name as the addressor.

Then he heard the familiar, loud, firm, authoritative voice that he knew so well. His mother said, "DAVID aren't you in that bed yet? GO TO BED. You have to get up in the morning."

"Like I don't know that." He thought but, didn't dare transpose the thought to audible words.

He went to bed without having completed the job of addressing the Valentine cards and decided that if he were lucky he'd be able to finish the task in school the next day. That was not to happen. He eventually learned that the only luck that can be considered to be reliable is always bad.

Mrs. Gaston assigned one of the students the responsibility for carrying her trash basket from beneath her desk along the rows of student's desks so that each student could deposit his or her cards in the basket which was then shaken to re-distribute the cards and two students were selected to "draw" and read the addressee and addressor's names on each card as they drew them. They would then hold the card while the addressee rose from his or her desk and walked to the front of the room to retrieve their card and convey their thanks and appreciation for the card to the addressor.

All went well for a few minutes and then the "drawing-distributing" team said, "To Joyce. Mrs. Gaston, that's all it says."

Mrs. Gaston asked, "Who's sending this card to who?"

David raised his hand and wondered, "Why the question about who it's to?" There's only one Joyce in the room, so he answered the question."

It wasn't long before a very similar scene presented itself and the same response from David was necessary. Mrs. Gaston was becoming slightly abrupt when making her comments.

The scene that is described in the preceding paragraphs was repeated approximately fifty times that afternoon. David decided that the two youngsters who made up the drawing-distributing team were deliberately pretending to not know whom many of the cards were being contributed to or from whom. Those two kids were not dumb or stupid. They were simply reveling in his misery and were not passing up the opportunity to cause him to suffer.

It didn't require many repetitions of the scene to cause David to develop a strong lack of appreciation for the celebration of St. Valentine's Day. He decided early that afternoon that he was looking forward to the time when all of his classmates matured enough to forego the nonsense he had to endure. He made the resolution to, if he lived long enough to get out of Evergreen City School, never buy another Valentine. When he reached Junior High School there was no formal celebration of that infamous holiday so, he didn't purchase any cards and when he began to receive them from his classmates he was forced to erase the notes on the one's written in pencil, re-address them, and redistribute them.

Remarkably, he never received any more Valentine's Day cards in the eighth through the twelfth grades.

After the sixth grade he didn't purchase another Valentine's Day card or gift until 1991 when he purchased them to present to the lady who became his second wife."

In 1946 and '47 he was in the fourth grade and considered to be old enough to carry a pocket-knife; some of the boys in the class also carried marbles, but he didn't. There were some boys carrying marbles who would play the game for, "keeps" where the winner would keep all of the little glass orbs that he knocked out of the ring and thereby increase his treasure as days and weeks went by. The adults in the school who of course, had all of the influence and could enforce rules issued the edict that playing marbles for, "keeps" was gambling therefore, the game was forbidden on the school ground. That only left the sanctioned game of baseball to play for entertainment during recess and the lunch breaks. David did not participate in that boring waste of time. The only thing the other boys would allow him to do in a baseball game was stand in the outfield waiting for something to happen. Remember? He wore glasses.

There was another game though that he could and did play with his pocketknife called, "mumbly peg". In later years when morality and any sense of personal responsibility and integrity were no longer considered to be important characteristics in one's psyche a boy would not be allowed to carry a pocket knife but, Tom Sawyer and Huckleberry Finn had carried Barlow knives in their day. David and his male classmates also carried pocketknives even though they were only eight and nine years old. He became a pretty good player of mumbly peg, which is a series of knife throwing movements.

As the year in the fourth grade progressed even mumbly peg became a bore and the teacher, Mrs. Freeman had become frustrated in her attempts to keep the class members from talking instead of doing their assignments. She started issuing detention sentences to those she caught talking or misbehaving in some manner. She started making David remain in the classroom during recess and lunch repeatedly writing sentences such as, "I will not talk in class," a prescribed number of times. The assignment varied in length according to the severity of the infraction and was intended to inspire better deportment and improve penmanship. It failed in achieving the first objective and his teachers quit scoring his penmanship.

In 1947 or '48 one of the three founders of one of the major industries in Evergreen named, "Southern Coach", Ceylon P. Strong purchased a J-3 Piper Cub and using material salvaged from a burnt building at the south end of his factory building on Pecan Street constructed a "T" hangar on the west side of the ramp at the airport on the site destined in later years to be the site of a large metal hangar painted red. Mr. Strong's hangar looked so bad that no hurricane, thunderstorm, lightening strike, or any other natural destructive phenomena would go close enough to damage it. He asked Mac to give him flying lessons and Jewel verbally assaulted him. He departed the McKenzie residence without having received flying lessons.

In September '47 he went to the fifth grade where Miss Lizabeth MacMillan, Mr. Mac's sister was his teacher and the only significant event he remembered from that year was the death of one of his classmates that he didn't know very well, Billy Morrison. Billy succumbed to natural causes.

One morning when entering the classroom David was surprised to discover an adult male in the room dressed in a suit of clothing obviously intended to be worn by a person who performed manual labor to financially support his family. "Pee Wee" Middleton was unusually quiet and sitting in his assigned place in the classroom without a smile.

Miss MacMillan entered the room and after unlocking her desk and arranging a few items on top of it she and the visitor left the room. She soon returned alone and announced that the visitor was "Pee Wee" Middleton's father who would be delivering "Pee Wee" to class every morning and retrieving him every afternoon because "Pee Wee" could no longer ride the bus to school. There were no reasons or explanations for the cause of that problem offered and of course, the students did not dare ask, "What's this all about?"

During the recess and lunch breaks "Pee Wee" seemed to be quiet. That was unusual but, his classmates respected his social withdrawal and in about a week the story began to slowly and piece by piece leak out of its, "lock box".

"Pee Wee" didn't like his school bus driver, Mr. Lambert and the sentiment was returned. "Pee Wee" was consistently searching for some form of retribution to apply to the bus driver for his attempts to inspire modification of "Pee Wee's" behavioral patterns and one afternoon while wandering around in the woods behind his home he discovered the perfect tool to apply to the problem.

When the U.S. Mail carrier delivered the mail to a rural address he'd simply place it in the mailbox and when the resident at that address wished to mail a note he/she would place it in the mailbox and raise the small red flag that's attached to the side of the box as a signal to the mail carrier that he should pick up the mail. "Pee Wee" discovered the carcass of a deceased, "pole cat" (a skunk.) that he retrieved somehow, and placed it in Mr. Lambert's mailbox on top of the outgoing mail.

Several days later Mr. Lambert asked the mail carrier why he wasn't providing service and was told to clean up the mailbox. One look inside the box after he overcame the odor surrounding it was the only clue needed to know who was responsible for the problem. He then visited "Red" Middleton and advised him that "Pee Wee" would no longer be allowed on his bus.

It was also the time when there was an attempt to form a Cub Scout group in Evergreen and Jewel volunteered to be a, "Den Mother". The position required her to host meetings of the, "Den" of Cubs who resided on Pecan Street in the McKenzie home and it became especially annoying to David when some of the Cub Scouts who met in the McKenzie home noticed that he had an impressive collection of comic books that he had accumulated at the rate of one per week from George Stamp's Magazine Stand on East Front Street. Every Friday afternoon when Mac would leave the shop and go to the A & P grocery store to pick up Jewel and a week's supply of groceries he'd take David and Jimmy along and leave them shopping for comic books in Stamp's store and retrieve both youngsters and two comic books after loading Jewel's purchase of groceries in the car. David was very possessive of his comic book collection and when "Whit" Brooks, Jack Black, and Randy White who later played professional football for the Kansas City Chiefs (?) wanted to borrow them he refused but, Jewel told him to let them "borrow" the books. He was correct in his thinking and judgment. He never saw the comic books again. That experience reinforced his confidence that his first impression and judgment of a person was probably correct.

The next year, the last four months of 1948 and first five months of 1949 were in the sixth and final grade at the Evergreen City School. The class was taught by Miss or Mrs. Cassel and the only things he remembered from that class was the substitute teacher, Mrs. Griffin who taught the class for several months that Miss or Mrs. Cassel was ill. It was just another year except that one day the entire class had to listen to someone who was not a teacher for a couple of hours and didn't understand what he was talking about but, at the end of the speaker's address discovered that they were now members of the 4-H Club. David never knew why, never understood it, and didn't care. He wasn't interested in farming.

During recess and lunch period the students who later became the "star" athletes in the class would play baseball and David would skate in the newly paved driveway on roller skates that Jewel and Mac had acquired for him. One of the youngsters in the class was Levon Shaver who was older than most of his classmates and was in the class because he could not meet the scholastic or academic requirements for advancement and the class, "caught up with him". He was promoted along with the rest of the class because of his age, three years older than the rest of the class.

Levon's classmates always approached him carefully. If they were alone, they treated him as an accepted equal but, sometimes groups of kids would be used as a protective shield by some kid who would walk by him when he was off guard or distracted and say, "Hi, Dap." Or, "Dip, Dap, Damn dope." They would then run as though their lives were in jeopardy with him close behind in pursuit. He never caught anyone because of not trying very hard. It's probably a credit to him that he could tolerate their brutal, inconsiderate teasing, and then frighten them as badly as he did without ever striking anyone.

The sixth grade students were mostly twelve and thirteen year old children except Levon. Then a sixteen-year old youngster was brought into the class that none of the students were acquainted with. Suddenly they were associated with a sixteen-year old contemporary who had left school in the sixth grade and started riding cross-country in a truck driven by his father. When he reached the age of sixteen he got a driver's license and began to drive trucks professionally at which point in his life the truant officer caught up with him.

The students tended to admire Rufus T. Lynch because he was one of them and simultaneously was already able to earn a living and they thought he was experienced and worldly wise. Several of the boys in the class wasted no time becoming, "his friends". They'd tease and harass Levon until he was compelled to chase them away and during the chase they'd lead him past Rufus where they'd accuse Levon of trying to, "beat 'em up" and plead for rescue by Rufus. Of course, they were trying to cause a fight to start and really didn't care which of the, "big boys" won. Regardless of who won the fight it would be entertaining to them.

No, David didn't think he learned much in class in the sixth grade but, he had learned something on the playground. He has since wondered whether or not it's part of the human psyche or a carefully planned tactic driven by selfishness to encourage confrontation between others and then hope that the confrontation will lead to incapacitation of the participants which will allow the instigator to inherit all of the benefits and property? He has seen such behavior consistently in society.

Neither David nor his classmates know what became of Rufus T. They never saw nor heard from him after they left the sixth grade. Levon remained with them until graduating from High School in 1955 and within a year drowned in the Gulf of Mexico.

In early June 1949 he was promoted out of the Evergreen City School along with his classmates and still has the Silver Dollar that Mr. Mac, the Principal gave him as a graduation gift.

PART X

JUNIOR HIGH SCHOOL

During 1949, '50, and '51 Mac would occasionally give David some, "dual instruction" in the Luscombe 8-A that he owned but, it wasn't serious instruction and undoubtedly, did not contribute to David's later accomplishments.

That Luscombe was the last airplane that Mac owned because of the pressure and resistance from Jewel. She had developed a strong and sincere hatred and distrust of pilots, airplane owners, and anyone else associated with aviation. She frequently expressed her negative and belittling views and opinions on the subject.

Since he was "laid back," and easy going Mac yielded to the demands from Jewel. He probably sensed that her objections to aviation had reached the level where continued efforts and expenditures in the activity endangered the family unity. He had made the decision to partially concede to her desires in 1942 when he quit instructing in the CPT program and returned home instead of continuing and moving the family to his employment site. No one knew how strongly he loved her.

Jewel's niece, Billie, one of three sisters that grew up in Mobile had married Ed Malloy who only had a, "motorsickle" for transportation became very friendly with Jewel and Mac and visited regularly.

Mac traded Ed a '40 Ford coupe that he had acquired for the Indian "motorsickle", disassembled, cleaned, repainted, and reassembled it. The bike became quite an inspiration for initiating debate and conversation with the customers who visited Mac's Repair Shop and had the audacity to ask Mac if he could ride a bike.

Jewel said she was, "standing at the sink preparing dinner, looked out the window and saw Mac standing on the saddle of *that motorcycle* with his arms spread out like the wings of some big stupid bird coming down the middle of the street."

She ran to the door, called the boys, put them in the house, then went to the shop where she interrupted the men who were acknowledging that they were incorrect in assuming that it wasn't possible to stand on the saddle of a motorcycle while riding it and congratulating Mac on the success of his demonstration.

Jewel proceeded to expound verbally on the wisdom of executing such a performance, the mentality displayed by a man who would do such a thing, the motives of men who would encourage such a spectacle, the lack of real potential for profit, how immature the exhibition was, the potential danger to innocent children in the neighborhood if Mac lost control of the vehicle or fell off, and that if the runaway-motorcycle ran over one of the kids in the neighborhood and killed it they never would get that paid for, and it would cost us everything we have, - - - - - - ." No one knows if she ran out of breath, became fatigued, frustrated, surrendered the idea of educating Mac, or simply gave up.

One by one the customers left the shop and the verbal tirade that Mac was suffering. Jewel vowed that she had her schedule full trying to take care of and control two boys and "M-A-A-A-A-C".

Yes, David attempted standing on the saddle of his bicycle but, was never able to release his hold on the handlebars and stand erect while going down the street.

Mac sold the, "motorsickle" to Wayne Congleton who resided in a house north of the McKenzie's on Pecan Street. Part of the payment that Mac received was a Servi-cycle, a very lightweight motorcycle that had once been owned by Bobby Johnson.

In September 1949 Jan and David's class entered Evergreen Junior High School where the class size expanded again because of the consolidation of two more groups of students joining them from Brownville and Annex, outlying communities in Conecuh County. By that time the pair of friends had grown to a threesome with the inclusion of Warren Lisenbee who was in the same class and had moved to Evergreen when his father became the manager of V. J. Elmore's dime store where one could obtain Valentines.

The teacher of David's section of the class was Miss Alene Robinson, a strict lady who no student was inclined to challenge. Her specialty seemed to be teaching English and she focused on diagramming sentences. That must have been an effective method for teaching the subject. While in that class he began to draw automobiles in his notebooks and textbooks. David has never regretted anything about his classroom experiences in the seventh grade. He also had his first exposure to any attempt at organized athletic activity. It was through participation in a class called, "Physical Education" taught by the High School football coach, Wendell Hart.

The class schedule kept the students in the classroom from 8:00 to 10:00, broke for the morning recess, resumed at 10:15, and at 11:00 the entire class went to the gymnasium which was nothing but a building to play basketball in, where, "P. E. Coach" Hart would handout six footballs and the class of boys would run down to the football field, split into six groups, "choose sides" to make-up twelve teams, and play "touch" football for thirty minutes. Then they'd run back "up the hill" to the gymnasium, return the footballs, and race across the school's lawn to the lunchroom which was a house that had been purchased by the Conecuh County Board of Education and was being used as the lunchroom. There was no time allocated for changing into gym suits, sweat pants, or anything else. There was no changing of clothing, no showers after the class, nothing different or special was thought of it.

At the end of the first six weeks grade-marking period David was given a "D" in "Physical Education".

Jewel "went ballistic" and David couldn't explain why he had received a "D". There was no written examination, his attendance was perfect, and he hadn't been late to class. He knew of nothing that he could have done incorrectly or "wrong" Jewel telephoned the Principal, Jack Finklea who promised to interview the coach who taught the class and then advise her of the reason for the poor score.

When Mr. Finklea returned the telephone call to Jewel he advised her that he had been unable to find a reason for the grade being so low and Coach Hart had agreed to revise it upward to a "C".

Jewel wasn't happy and David was frustrated, insulted, and absolutely angry. From that time onward throughout his life he considered participants in the physical-athletic sports to be a collection of self-serving, egomaniacal examples of a "good ol' boy's fraternity" and he always referred to them as the, "damned jock strappers". In later years David, like everyone else, had an occasional encounter with a, "damned jock strapper" and always had to suppress the warning flag that sprang up in his mind the instant that he became aware of the other party's association with sports.

He must have been intimidated by the older kids in the school building because the Junior High and the High School shared the same building and playground making the total attendance about three hundred seventy five students. That meant he was twelve to thirteen years old in the seventh grade and the average High School Senior was eighteen years old. Jan Hendrix quoted David as having said, "I feel like I'm back in the first grade."

That was the summer that Mac traded the Servi-cycle to George Hendricks for a homemade motor scooter that he allowed David to have. Until that time David only had a bicycle that he had been given for Christmas when he was in the third grade and Jewel was very concerned about him being injured in traffic on it. She didn't display the same concern about the motor scooter. It seemed that every time David left home on it he pushed it back when there was a failure of some part on it, usually on the engine.

That summer Jimmy fell seriously ill again. The symptoms were the same in 1950 as they were in '47 so, he was hospitalized in Greenville again and the diagnosis was, mononucleosis and he was returned home and put to bed for most of the summer.

Jimmy had saved almost thirty dollars by this time and Mac found a 1926 Model "T" Ford that Jimmy bought for twenty dollars. Imagine a nine-year old child owning an automobile that was drivable. Of course, he never drove it but, it was undoubtedly, a very good psychological ploy exercised by Mac to boost his morale. Jimmy sold the car at the end of that summer for a greater price than he paid for it. The ability to sell for more than he paid became another passion to him.

While Jimmy was hospitalized Mac discovered a "Hobby Shop" in Greenville and perhaps twice per week he'd visit that store when he went to visit Jimmy and on one of those occasions he bought a .074 cubic inch displacement, two cycle, model airplane engine, and a model airplane kit. Jimmy never displayed any interest in either one but, later that year David certainly did.

Mac had assembled the model airplane but, since he never used paint, wax, a mop, or a broom the model airplane constructed of balsa wood was, unfinished. The wood wasn't sealed which led to it absorbing more fuel than it consumed and its weight probably tripled. It never flew.

"Air Trails" magazine was delivered to 217 Pecan Street every month and in late 1950 there was an advertisement in the publication for a model airplane produced by the A. J. Walker Co. called the, "A.J. Firebaby" that proclaimed the kit to provide everything needed to fly including flying instructions. The price without the engine was three dollars and fifty cents so, David ordered one.

Most youngsters receive an allowance. David's was fifty cents per week and he must have used all of that and whatever he had saved to purchase the "Firebaby" while Jewel was probably saying, "That's just like a pilot. He'll spend everything he has on a damned airplane."

He was promoted to the eighth grade, which was divided into two sections. One was Mrs. Millsap's homeroom and the other homeroom belonged to Mrs. Shaver who's brother-in-law owned the Post Office in Herbert, Alabama that was also a small, "Country Store" that had been mentioned in Robert E. Ripley's column, "Believe it or Not" because of the sign on the front porch of the dilapidated building that sagged in the middle stating, "Absolutely No Business on Sunday and Damned Little During the Week". It is debatable whether Mr. Shaver was espousing the sanctity of the Sabbath, complaining about the economics of the area, or both. Mrs. Shaver never mentioned her brother-in-law to the eighth grade class.

She taught English and History and Mrs. Millsap taught Math and Science. It was in the science class that David wrote a paper about flying model airplanes and he offered to demonstrate a flight for the science class if he were asked so, one morning both sections of the eighth grade met at the football field where he performed the, "exhibition flight in the name of science." The exhibition was successful and he was rewarded with applause from the other students, a good grade on the paper from

Mrs. Millsap, and challenged to control his ego. The high school, "grapevine" spread the news of the model airplane flight demonstration quickly and by the end of the day David's social circle had expanded from three to six, seven, or maybe even eight. They included Mr. Eugene Patten, his sons Joe and John, Allard French, Glenn Pate, and John Wilson. He may have been the first person in Conecuh County to successfully fly a "U" controlled model airplane. It's remarkable how success or failure affects one's social standing.

Eddie Tuggle was the most talented artist in the class and he made a cartoon of one of the teachers that depicted her as a Communist Country's Army Officer. Her response when she accidentally saw the cartoon totally eliminated any chance of any of the students continuing to appreciate the humor they had been finding in the illustration. Eddie became probably the only male cheerleader that Evergreen High School ever had and within a year of graduation drowned while water skiing in Tampa Bay.

That was the same year that one of the older students in the class decided to take a joy ride in a new bus that he "liberated" from the storage lot of its manufacturer, "Southern Coach Manufacturing Co.", much to the amusement of the other students. The local gendarmes and judiciary dealt with the consequences of that episode.

The summer of 1951 was a very busy one for David but, he didn't realize that he was involved in so many activities or was being introduced to so much experience in life. That was the summer that he started working part time, some of the time for his father. His job schedule wasn't strict at all. His duties were to remove radiators from vehicles that came into, Mac's Repair Shop for repair in the morning, place them in the cleaning vat, place scrap wood beneath the vat, and start the fire. Then he seemed to be free until 3:00 o'clock that afternoon when Mac and Hugh would have the day's order of radiator repairs completed and ready for reinstallation into the vehicles, which David assisted in doing. He didn't object to working on Chevrolet radiator removal and installation but, Fords were a different story. They were larger, heavier, had twice as many hose connections, and sometimes had cooling fan shrouds to wrestle with. He thought he was earning the five dollars per week that Mac had begun to pay him.

During that summer John Wilson purchased a "Firebaby" kit and would regularly visit David at Mac's Repair Shop from where the two would cross the street to the empty lot in front of Mr. Strong's factory building, set up the models and take turns flying them. The customers at Mac's shop would stand in front of the shop in the parking lot and watch the model airplane flying while awaiting their turn for service from Mac or his employee, Hugh Mason.

No other business in Evergreen could claim to present entertainment for its waiting customers and neither the customer nor Mac's Repair Shop had to make a financial contribution for the service. They didn't either.

It was almost obligatory for a young male who desired social acceptance in Evergreen to join the, Boy Scouts of America where it appeared that the primary interest and skill encouraged and developed was swimming. They met every Monday evening at the Evergreen City School and there is little doubt that W. Paul "Mr. Mac" McMillan, being the Scoutmaster had something to do with that. David had encountered no difficulty moving up in rank from, Tenderfoot, to Second Class, to First Class, and then qualifying for, Star but, even though he had earned the five Merit Badges required for the rank or grade he was never awarded the goal. He never asked if his qualification for promotion had been presented to the, Board of Review thus, never knew if it had or not. He had also been the, Troop Scribe for the past several months and that job came easily.

Every Tuesday he'd write a short report on the events that occurred in the Scout meeting on Monday evening and leave the copy at the local newspaper office, "The Evergreen Courant" and they'd publish it every Wednesday which may have been good ego fodder for him but, he had already decided there was very little competition for the news copy space in The Evergreen Courant.

Every summer the scouts would board the small school bus that the Troop owned and go to Camp Big Heart located east of Pensacola, Florida on the mainland side of Santa Rosa Sound for one week that he expected to be like a vacation.

One Sunday morning the scouts going to camp for the week boarded the, Scout Bus where it was always parked on the playground in front of the, "City School". One of the older scouts drove his own car filled with friends and the other thirty or thirty-five boys boarded the, Scout Bus. Mr. Mac engaged the starter of the bus and it surprised David when the engine started one more time. He always expected the bus to fail sometime, somewhere but it always performed its duty.

Pensacola is about seventy five miles southwest of Evergreen and it required about two hours to complete the trip while the scouts tried to pass the time by singing the song, "One Hundred Bottles of Beer on the Wall" and repeatedly noticed that the older scout's car kept passing them. They occasionally saw the car parked at a country store or service station and knew that it wasn't there for gasoline.

Upon arrival at Camp Big Heart David noticed that there was very little grass, just small trees and brush growing out of the sandy soil and all of the buildings may have been described as rustic but, it appeared to him that they were really crudely constructed by amateurs and weren't worth painting. The cabins that they stayed in were wooden framed with surplus army bunks for beds and the entire assemblage then covered by an army tent. Yes, it was "camp". The scouts ate supper in the, "mess hall", listened to the announcements of schedules, expected behavior standards, and camp procedures. They were then dismissed and went to their assigned cabins and climbed into their selected bunks. David thought, "The army must be similar to this."

Monday morning began with the, "wake up call", a visit to the latrine, as they called the community bath house and toilet, made up their bunks, walked to the, mess hall, ate breakfast, returned to the cabin, "policed" the area, and changed into bathing suits. They then went as a group to the end of the pier that extended into Santa Rosa Sound for the morning's swimming activities. From the swimming tests and lessons the scouts went to another activity and then to dinner at the, mess hall. That afternoon they repeated the same schedule. David did not consider the activities to be useful at all. The only thing they seemed to do was consume time and demand submission to the commands of adults who believed they were providing essential guidance and education. Those adults were also asking the scouts to work on something constantly so, he selected two Merit Badges to attempt earning, Aviation and Machinery. There were no counselors who were qualified to judge him on those subjects and they had another excuse for not allowing him to attempt to earn the one for Archery. He was becoming disillusioned, frustrated, and disgusted with all of it.

All the older scouts had a cabin for themselves and appeared to be aloof, indifferent about, and excluded from the activities required and demanded of the younger scouts. They didn't even seem to be noticed by the staff and counselors. The younger scouts never saw any of them at any of the sanctioned activities but, became acutely aware of their presence when they were singled out and one by one asked, "Who's army are you in, General Jones's or General Smith's?" by an older scout.

The younger scout was then faced with having to correctly decide whether his interrogator was a friend of Smith or Jones and hope that he made the correct choice when he answered. The decision was complicated further by the obvious fact that all of the older scouts seemed to be equally friendly with each other and the interrogator was likely to pretend that he liked or disliked either Smith or Jones and there were no rules to guide him in making the choice. The younger scout was in a dilemma.

The camp staff and counselors were most likely unaware of the establishment and organization of two, "armies" and certainly didn't notice their existence as the younger scouts called, "Privates" were commanded by the two "Generals" or their "Staff Members" of lesser rank, but no lower rank than "Colonel" to yield to their demands which were usually to perform some menial chore.

On Thursday morning while the staff and counselors of the camp were in a meeting the two, "Generals" ordered all members of both, "armies" into their cabin and had them sit side by side on the lower bunks around the perimeter of the structure. They posted "lookouts" even though the where-a-bouts of the staff and counselors was known and ordered all of the, "Privates" to loosen their belts, unbutton or unzip their trousers fly, expose their masculine genitalia, and proceed to masturbate.

A few, very few, perhaps three of the younger scouts called, "Privates" refused and the older boys promptly removed them from the, cabin which left the impression among the other, "Privates" that they had better comply or else they would become victims of some unknown and terrible consequence as a result of their insubordination. That was a lesson in the effectiveness of mass psychology.

One, "Private" was physically mature enough to achieve ejaculation which seemed to please all of the, "Generals and Colonels" who then dismissed all of the, "Privates" without bothering to admonish them to never relate the experience they had endured to any one. It was probably the combination of pride and shame that prevented any of the younger boys from ever mentioning the incident, even to fellow participants. The staff and counselors at Camp Big Heart most likely were never aware of it.

That Saturday the members of Troop 40, BSA except the five boys who rode together in the automobile boarded the rusty, battered Scout Bus and returned to Evergreen where, upon arrival word spread quickly, all over town that the, "scouts are back", and parents, friends, and relatives picked each boy up and asked, "Did you have a good time?"

Yes." Each boy lied, and that was the last time that David went to Camp Big Heart.

PART XI

LIFE & ENVIRONMENT CHANGES

In August 1951 the cotton gin located on the west side of and diagonally across the street from the McKenzie home on Pecan Street opened for the annual ginning season. When cotton ginning has been mentioned in books and shown in movies it was presented as colorful, "folksy", and a time of relaxation or celebration of the completion of a working season. No one has ever thought cotton ginning could possibly have any objectionable features associated with it. The residents on Pecan Street in the fall of 1951 probably felt that way but, had change forced and imposed upon them by the operation of the Miller Gin Company. Jewel and Mac had never expected to encounter the lack of consideration, selfishness, and indifference such as they were about to discover in such a quiet small town.

Although there were mechanical cotton pickers in existence in 1951 and were used in the south they were not commonly seen in south Alabama because of their cost and shortcomings. It shouldn't be necessary to point out that their cost was prohibitive for the average family sized farming operation and the machines could only be used one time per season in a cotton field because they destroyed it. They operated by stripping the cotton plant of leaves, twigs, branches, the shells from the opened cotton bolls, cotton that was ready to harvest, cotton that was not ready for harvesting, everything, and anything above ground level. The use of the machines had to be carefully selected and controlled because the farmer wished to delay the entry into the field until as much cotton as possible was ready for harvesting and if he waited too long, it might rain which would cause the open cotton to be stuck in the boll as though it had been glued in place. Because of the reasons listed the farmer dared not pick cotton too early with a machine. He would have to face the potential elimination of his future profit. He was financially depressed if he did and impoverished if he didn't.

Hand picking was the accepted method of harvesting the cotton crop in south Alabama because it could be performed selectively and repeatedly allowing the farmer to harvest everything a cotton plant could produce and the labor cost was not prohibitive, ten cents per pound. A person who practiced the temporary and seasonal profession of, "cotton-pickin'" could pick a hundred pounds per day by, "hustlin'". Think of it. That's ten dollars per day, six days per week, four and a third weeks per month, for half of August, all of September, and most of October. Why, that's the potential for six hundred forty five dollars to a person with no skill and no professional costs accompanying them.

Those are also the reasons that a cotton gin had to operate from mid August until approximately the first of November. The gin owner might justify an expenditure on the installation of cotton cleaning equipment because a cleaner bale of cotton sold at a higher price than one contaminated with parts of stalks and leaves. The gin owner was also led into seeking higher profit by adding the cleaning equipment instead of budgeting for maintenance.

The gin began operations in the morning and starting time varied according to the expected demand that was dictated by the time that farmers started forming a waiting line at the site. If demand was low and the waiting line short the gin might not be started until 9:00 o'clock in the morning or, if the lines started forming early it might be started at 5:45 in the morning but, in any case it ran until there were no more customers waiting for service. Some evenings that was as late as 11:00 o'clock in the evening but, it was a seasonal operation and that made all of it acceptable until the installation of the cotton cleaner and the lack of maintenance on the engine and its noise suppression system combined to create a nightmare, a nuisance, and an environmental abomination.

In August 1951 the diesel engine powering the gin started up slowly just as it had done every year before but, this time it was unbelievably loud and as it gained speed the sound of the newly installed cleaner joined in arousing everyone in the neighborhood and when they looked in the direction of the gin they saw the oily black smoke coming from the end of the exhaust stack.

Very soon after the gin started it began to eject dust, dirt, and cotton lint into the atmosphere as each vehicle in the line of trucks, tractors, and mules pulling trailers and wagons took their turn to park beneath the pipe that was the pick-up tube operating like a vacuum cleaner. Just as happened in prior years the gin was soon operating from 6:00 o'clock in the morning to 11:00 o'clock in the evening and with no muffler on the seriously worn diesel engine's exhaust and no filter on the cleaner exhaust it required only a week or two to have an oily deposit on everything in the neighborhood.

Children couldn't play on the lawns. Nothing outdoors could be kept clean or used. The window fans in the homes had to be installed to exhaust the houses and the cotton lint, oil soaked of course, was collecting on the rest of the window screens in every house in the neighborhood. Naturally, ordinary window screens couldn't possibly filter all of the contaminants out of the air going into the homes occupied by residents of the neighborhood and no one was conscious of environmental hazards to public health or if they were the owner of Miller's Ginning Company and members of the City government chose not to exhibit any concern.

It wasn't long before the entire neighborhood was upset, concerned, and desiring that corrective actions be initiated. Jewel, Mac, and Bertha and J. C. Johnson became the leaders who were designated to contact the owner of the gin and discuss the situation, which lead to no relief being offered. They contacted the City Officials and asked for help, but they encountered the, "good-ol' boy" blockade and received nothing except smiles, frowns, statements of sympathy, and expressions of incredulity that they were so audacious as to complain about the gin owner's disregard for the quality of life and neighbor's property values on Pecan Street.

Of course it would have lead to financial hardship for a lot of people if the gin were shut down or closed so, no action was taken to alleviate the complaints and the residents were forced to attempt to continue living in totally unacceptable conditions. The air was filled with dirt, cotton lint, and noise, which the City Officials chose to disregard so, Jewel, Mac, the Johnson's and a few others started a petition drive, hired legal assistance, complained at the State Government level, and conducted a fight that lasted approximately two years.

The gin owner overhauled the diesel engine, installed a muffler on the exhaust, and installed air-cleaning equipment on the cotton cleaner before beginning operations the next year.

A few years later the gin closed. Jewel and Mac had become known as troublemakers in the "good-ol' boy" network even though the hazardous nuisance had closed due to economic changes in south Alabama. The closure of the gin was not due to any consideration of the health of neighbors or their life styles and living conditions by the political forces that ran the community.

The population count in Evergreen, approximately three thousand five hundred justified having five City Councilmen and a Mayor. Four of the five Councilmen were also Elders or Deacons in the Presbyterian Church. Mac was also a Deacon in the same church. Jewel was offended by the hypocrisy that in her view was exhibited by those four Elders and Deacons who were also City Councilmen that had displayed complete indifference to the living conditions of other people and still professed to be upstanding members of the community and the Church.

The McKenzie family ceased to attend the church.

Jewel and Mac led an effort to establish a zoning ordinance in Evergreen, which had never had one. Even today its appearance will cause an observer to wonder if there is one in the community.

In 1951 Mac ran for a seat on the City Council but, was not elected and in September of that year David began the school year as a High School Freshman in a homeroom made up entirely of boys with the Diversified Occupations teacher, Mr. Elmer Taylor charged with monitoring the room, supervising, and attempting to control the boys. The "Homeroom" session lasted forty five minutes to begin the school day after which the students scattered to their respective class assignments but, while the boys were together in Mr. Taylor's homeroom pandemonium reigned supreme and the teachers in the other rooms on the same hallway closed the doors to their classrooms.

Mrs. Walter Lee taught the only class that David enjoyed that year, English and Literature. Mrs. Lee exhibited the professional sophistication to allow her to sit in a "desk chair" like the students were using placed in front of the class. Not behind a desk as though it were a shield. She had an aura about her that caused the students to feel that she was interested in their welfare and liked them but, they also felt that she probably would not tolerate any challenges and she had the leadership ability to project that image without appearing to be deliberately attempting to do so. David would never forget that it was Mrs. Lee who taught the class a small amount of etiquette and some of the social graces. He always wished that she had taught more. As it was, she was the only teacher he ever met who he thought tried to do that and he always wished that he had more exposure to it. He liked Mrs. Lee.

Even though he was actively involved in building and flying model airplanes he was convinced that he'd never fly so, he sought an alternative goal. The search was not deliberate or planned but his home environment drove the decision. Hindsight whose acuity is always twenty-twenty shows that his choice was almost predestined. He hoped to become a racing car driver and his alternate choice was to become an automobile designer. He had been making sketches of automobiles in his notebooks and textbook covers since he was in the seventh grade. His interests began to intensify and specialize in automobiles.

During their last fund raising campaign the senior class had been selling magazine subscriptions to provide funding to the school and to the senior class to assist in financing some of their graduation celebrations and David bought subscriptions to, "Motor Trend" and "Hot Rod". When the family would visit Mobile or Montgomery he'd make it an objective to visit the newsstands that stocked magazines scouring the sections devoted to and specializing in automobiles. He was very protective of his collection of magazines specializing in automobiles and had never forgotten the disposition of his comic book collection when he was in the fifth grade.

He still had the motor scooter and his bicycle but, since he had begun to move automobiles and trucks around Mac's Repair Shop his interests were being diverted, not only because of Jewel's opposition to aviation, but also due to the availability of an alternative.

During the Thanksgiving holiday that year, 1951 David rode with Hugh to pick-up Mac after he delivered an automobile to a customer and observed a negotiation that affected the family from that day onward. The car they were delivering belonged to an African-American and normally Mac wouldn't have closed the shop and had his employee, Hugh help him deliver a customer's car but, this customer was a "regular" and good one. Usually the customers were required to bring their own jobs to the shop and if they didn't wait for job completion they'd have to return and retrieve it later. Mac did not pick-up or deliver but, this was an exception.

At that time in south Alabama an African-American wasn't expected to approach a Caucasian's front door. They'd go around to the back and knock on the rear door when they had occasion to visit the home of a member of the "majority" race and when a Caucasian approached the home of a member of the "minority" race he would not step onto the front porch. He'd remain on the front walk or most often on the ground at the steps and knock on the floor, steps, or whatever was available to attract the occupant's attention. Negotiations or entry occurred after the initial approach and recognition was completed and in this case Mac presented the keys to the vehicle, accepted the payment for the repair, and then asked, "Whose old clock is that under the steps in the dirt?"

"Oh, Mistah Mac, dat's jus' sum'n' de kids play wif. You wan' it? You kin have it."

Mac asked him to find all the kids and when they were all there he negotiated a purchase price for the clock divisible by the number of children and he became the owner of what was later learned to be an 1873 Seth-Thomas calendar clock. He didn't mention it to Jewel.

By the time Christmas arrived he had completely restored the clock except for the lower face and moved it into the house. Jewel didn't seem to believe that it was anything special so, it was placed on the mantle in the boy's bedroom where it remained until one of her friends saw it and said, "Ooooh Jewel, if I had that it'd be in my living room."

When Mac entered the house at 5:00 o'clock that afternoon he noticed that the clock wasn't in the boy's room and when asked about it, Jewel told him it was in the living room.

Mac leveled the clock on the mantle in the living room and Jewel decided that none of the furniture matched it. The solution to that problem was to start visiting antique shops whenever they visited any larger town and over the years she became known by antique furniture dealers from Pensacola to Birmingham and New Orleans to Atlanta and of course, every piece that was added in the house caused the replacement of another piece of furniture near it until all of the furniture in the house had been replaced with antique furniture. Then the process started again and that house and the next one that they owned went through four cycles of refurnishing. She had a respectable collection.

Mac's customers learned of the clock and what was happening as a result of its arrival and occasionally one of them would drive up to the shop, get out of their vehicle, and open negotiations by saying something like, "Mac, I need to have this - - - - - - - fixed and I have this old clock. Will you fix it in exchange for the clock?" Mac usually did.

That business practice led to the attic in the shop being filled with clocks, which he'd repair during the year and every Christmas some of the relatives and friends would receive an old clock for a gift. The family members never heard of any of those gifts inspiring a remodeling or refurnishing a home, though.

Mac always wanted a "Grandfather" clock and during conversations with customers who came into the shop he'd sometimes be told about the possible location of one which would lead to drives in pursuit of the rumor on Sunday afternoons. There were also attempts to track down old cars for Jimmy's sake. Those drives with the family were pleasant and Mac seemed to have a gift for meeting and talking to people.

PART XII

JIMMY GETS HIS OLD CAR

The clock led to becoming friends with an elderly gentleman, Mr. Daniel Shell, who was a "clock maker". At least that's what watch and clock repairmen were called, and Mac spent a lot of time with him. Beneath an oak tree beside the house was a Model T Ford that Mr. Shell bought new in 1923. The car had been parked there since 1925 when someone attempted to overhaul the engine and re-assembled it incorrectly leading to the number four connecting rod coming loose and going through the left side of the cylinder block when one of the nuts retaining the rod cap came loose. The engine and transmission assembly had then been removed from the car and stored in a barn.

Jimmy was eleven years old at the time and fell in love with the "23 T" so, Mr. Shell sold it to him, complete with the engine and transmission assembly from the barn for five dollars. On July 4, 1952 it was loaded on a trailer owned by Mac and hauled to Mac's Repair Shop.

And it was all because of a clock.

There really hadn't been anything memorable about David's year in the ninth grade. He had been one more of the students at Evergreen High School and his scholastic record reflected a lack of interest in the curriculum except for English and Literature. That was due to Mrs. Lee. He was undoubtedly skeptical about having a future in his newly chosen career as a racing car driver and couldn't see opportunity for anything in Evergreen. The A. J. Firebaby model airplane had been destroyed in an accident, specifically attempting to perform a loop with it and David had begun to experiment with ideas and designs of his own for models. Most of them were failures but, he was learning from the attempts regardless of the frustrations. The motor scooter was completely worn out, he had no interest in the Boy Scouts or athletic events and there weren't any social activities that he was aware of that were suitable for a fifteen year old boy in Evergreen. He did go to football games on Friday nights in September, October, and November, and the basketball games during January, February, and March but, attendance at those events certainly was not due to any interest in them on his part. He was simply being sociable and those were the only places where he was likely to find other youngsters. He was simply doing what he was told to do, no more, no less, and at the end of the summer he began his sophomore year in High School.

Mr. Howell was his homeroom teacher and he certainly wasn't like Mr. Taylor. The room full of students was co-educational and its occupants were quiet and controlled or as nearly so as can be expected of High School Sophomores. Dave probably thought that he had no achievable goal in his ambitions and, therefore he had no desire to apply any effort to the academics as the school year began and at the end of the first six weeks' grading period the report on his progress wasn't a good one.

Jewel's response to the poor report began vocally after he arrived at home that afternoon and presented the report card on demand, of course. He found escape by changing clothes and going to the shop next door to the house, but at five o'clock when he closed the business and returned to the house Jewel resumed her critique of his lack of academic achievement. The more she questioned, critiqued, and complained, the more frustrated she became until she exploded into a rage which was accompanied by a physical assault on the boy.

Mac, upon hearing the disturbance entered the room to investigate and he soon removed his belt and joined in issuing corporal punishment after which the entire family sat down to supper.

After the family attended a movie at the Pix Theater David was preparing for bed and found that the dried blood in several places on his back held his undershirt in place.

His grades improved.

During the Christmas holidays of 1952 Mac disassembled the Model "T" engine and transmission for "Jimmy's Model T", welded a patch over the hole in the left side of the cylinder block, acquired the needed replacement parts from Mr. Shaver's Store in Herbert, but not on Sunday, overhauled the engine, and then did some design, engineering, and modifications of his own.

All of the original parts for the ignition system had been misplaced and lost during the twenty seven years that the vehicle sat disassembled and unused and Mac had no interest in accurately restoring the car, so he retrieved a high tension induction type magneto manufactured by the Case Company for use on the eighty five horsepower, four cylinder airplane engine from his collection of parts in the shop. Then he modified the drive gear on the magneto and mounted the unit in the same location that the generator had originally occupied on the Model "T" engine.

Naturally, the wooden-spoked wheels had decayed while the car sat immobile beneath the oak tree for twenty seven years, so Mac retrieved four sixteen inch steel wheels from the same scrap pile that contained the remains of the "Jimmy-Wayne Flying Machine" and welded them to the steel hubs from the Ford's wooden wheels.

On the afternoon of Christmas Day, 1952 while David sat in the house peering out the window at the proceedings Mac and Jimmy started the engine, ran it for a few minutes, stopped it, let it cool, and repeated the cycle several times. Mac eventually drove it down the street and back with Jimmy in it, "sitting tall".

David was confined to the bedroom that day because he had contracted a disease called, "Roseolla", which was similar to measles, but of much shorter duration. Until Christmas Day he had been in the shop watching and assisting Mac. The Model "T" was then stored in the shed some times called, "a barn" and at other times called a, "garage" behind the home.

In January, 1953 David acquired a Learner's Permit, enrolled in the Driver's Education course at school, and started preparing for the driver's test for a license to operate a motor driven vehicle. After twelve lessons he was recommended for, took, and passed the test for the Alabama Motor Vehicle Operator's license. He had wanted to obtain the license on his birthday, the youngest age that a person in Alabama could obtain a driver's license, but the examiner from the Highway Patrol, they're called, "State Police" in most states, only came to Evergreen on two Thursdays per month and an applicant was required to be sixteen years old. He was sixteen years and one week old when he got his license and thought that was the greatest achievement he would ever experience.

The instructors had been Coach Wendell Hart who was giving the grade of "B" in Physical Education courses every six weeks and the Assistant Coach, Ralph Law. Of course, David did not tell them what he thought of, "Jock Strappers". Coach Ralph Law had come to Evergreen High School after graduating from college and soon became attracted to a daughter, Patti of Combie Snowden, one of the barbers living on Pecan Street.

Jimmy's 1923 Model T with David driving, Evelyn Cannon, Nell Cannon & Jan Hendrix in the back seat. The toddler is a cousin of the Cannon twins. (author's collection)

Case magneto from an 85 H.P. Continental airplane engine installed on the '23 Model T Ford engine by "Mac". (author's collection)

Patti Snowden had a cousin named Jimmy Snowden who was three or four years older than David and was noticeable because he displayed an ability to locate and purchase Model "A" Ford automobiles regularly for a maximum price of twenty five dollars and David thought that several of those cars were good candidates for "Hot-Rod" material. Jimmy Snowden was successful in maintaining employment with the merchants in downtown Evergreen and he'd purchase a Model "A" to drive for approximately a month, then paint it a bright color with a paint brush after performing no preparation for the new paint job. He'd paint a car in one afternoon. In another two weeks the fenders and running boards would disappear and in another two weeks the car would disappear and be replaced by another one, which was destined to follow the same program.

Stores in Evergreen always closed on Sunday and for one half day on Wednesday afternoon. Snowden was employed on a part time basis in one of the stores in downtown Evergreen and he was seriously interested in the game of golf. Every Wednesday afternoon he would get in his latest Model "A" and drive out to the Evergreen Country Club to play and socialize with the town's elite and it was during one of those Wednesday afternoon drives to the golf course that he noticed that the engine was beginning to over-heat in his latest car that was a coupe that had reached the fender-less stage. He also became aware of a need to alleviate the discomfort caused by the contents of his urinary bladder approaching the maximum capacity that the organ could contain. He decided that he should stop on the shoulder of the highway and solve both problems with one efficient operation.

The Model "A" Ford engine is an inline, four cylinder, "L" head, liquid cooled engine with the ignition distributor mounted on top of the engine between cylinders number two and three. The electrical current that causes the spark at the appropriate time in each cylinder is conducted from the distributor to the four spark plugs by unshielded copper strips. They conduct a very low amperage electrical current of twenty thousand to twenty eight thousand volts from the distributor to each spark plug.

When Snowden proceeded to cool the overheating engine and relieve his personal discomfort by transferring the contents of his bladder onto the running engine he learned that urine is an ionized solution and as such, is a wonderful conductor of electricity. Imagine twenty to twenty eight thousand volts of electricity traveling up a stream of urine and into the tip of his little "dingus". It did indeed "titillate his sensibilities."

When David heard the story of Jimmy Snowden's method for correcting mechanical malfunctions he lost any interest that he may ever have had in any of the cars that Jimmy Snowden may have owned.

Model "A" Ford engine Note the copper strips between the distributor and spark plugs (author's collection)

Jewel and Mac were very careful about allowing David the use of the family car, a '49 Ford tudor sedan that Mac had purchased from an insurance company when it was a total wreck with seven hundred miles on it for five hundred dollars and repaired it. They were probably concerned about David finding more trouble than he could handle or they could financially resolve. It definitely would not have required much to exceed the latter and little more to overcome his capability. Jewel expressed more concern for potential problems resulting from David's driving than Mac did, but David was aware that his father didn't trust him with the car. He was allowed soon after receiving his driver's license to drive the Model "T" to and from school and to use the vehicle almost as though it was his own car on Sunday afternoons.

David kept wishing the Model "T" had lights so that he could use it in the evenings, but Mac in very subtle ways kept him from ever installing lights on it. It didn't have a battery, generator, or starter and

David was being paid only five dollars per week to work part time for his father in the shop, therefore he couldn't financially support improvements in the car that really belonged to his brother, Jimmy.

At the beginning of the summer David started opening Mac's Repair Shop every morning at 8:00 o'clock and he at least, was present on the job eight hours per day. Hugh, Mac's employee had left and was employed by the County. Mac maintained his regular and long practiced working schedule, 10:00 to 12:00 and 2:00 to 5:00 o'clock. David didn't notice what his brother, Jimmy, was doing, and he took for granted and accepted the fact that his mother, Jewel was continuing as her usual industrious, meticulous, quiet, refined self except when she was annoyed or angered and absolutely no one dared attempt to annoy or anger, "Miss Jewel". That was a well-known fact.

On Sunday afternoons David would dress in a white shirt, dress slacks, sport coat, and wearing a Kentucky Colonel bow tie because he liked the appearance it presented when blowing in the wind created by the "fast moving Model "T"", go to the movie at the Pix Theater.

After the movie he'd drive around town picking up other schoolmates that he knew and then stop at one of the local gas stations, take up a collection from his passengers and purchase that amount of fuel for the car. It had a ten-gallon fuel capacity and during the time that he operated the vehicle it probably never had more than seven gallons in the tank. Gasoline cost nineteen cents per gallon at the Spotlight Drive In.

Yes, it not only may seem to be a rather boring way to spend the teenage years, but it was boring and there was nothing he could do about it. It may have been to his advantage or relief that he was bashful and shy because he did not enjoy the financial capability of supporting very much social activity and his interest in high performance automobiles coupled with an undisciplined inclination to test the capability of a vehicle justified the hesitancy exhibited by Jewel and Mac to allow him the use of the family car.

He entered the junior year of High School in September 1953 and became acquainted with Homeroom, Science, English, and History teacher, J. H. "Harry" Dey, Jr. whom he came to both like and respect very much. Mr. Dey was a bachelor and didn't seem to have any interests other than teaching and especially history which he presented in an entertaining, fast moving, real life, and interest arousing manner that David believed totally discredited any students' possible complaint about having to take the course. He remembers Mr. Dey often using the phrase, "and those old boys then". The science course that year was, Chemistry and Mr. Dey presented the course at a rate he adjusted to the student's rate of absorption and level of comprehension of the information presented. Mr. Dey was good.

David and all of the model airplane enthusiasts had gone their separate directions, which left his social circle at one, Jan Hendrix. Ray, Herston, and "Pee Wee" were in the same class, made up their own "threesome", and pursued their own interests. All three of them were called, "D. O. Students". They were studying Diversified Occupations and "Pee Wee" held a job at Fairview working for O. B. Salter's Grocery Store and gas station as a clerk, butcher, shelf stocker, floor sweeper, and gas station attendant. Ray was employed at the Conecuh Quick Freeze, a company that rented lockers to customers that didn't have a food freezer in their own homes, and Herston worked as a shoe shine boy, window washer, floor sweeper, and general helper at Snowden's Barber Shop in downtown Evergreen.

O. B. Salter didn't want his employee to play the traditional game called, "trick or treat" on him, so he chose to attempt to bribe "Pee Wee" when Halloween, 1953 arrived by giving him some of the only thing he had, several bags of groceries which "Pee Wee" shared with his buddies, Ray, and Herston. They didn't know what to do with groceries on Halloween.

"Pee Wee" had been presented with a bag of potatoes the preceding year and had spent the evening jamming the potatoes into the exhaust pipes of Pix Theater patrons' cars that were parked on the two streets in downtown Evergreen. Unfortunately, after he had stuffed all except two or three cars' tailpipes the local cops arrived.

The City Police were quite capable of meeting the challenge presented by the perpetrator of the dastardly deed of, "tailpipe stuffing" since they had qualified for employment on the police force by successfully serving for six months as drivers for Simp's Taxi. They coerced or sentenced "Pee Wee" to go from car to car removing potatoes while they explained to each driver the reason for his or her car failing to start and run. "Pee

Wee" didn't care to repeat that experience, but what was he to do with groceries on Halloween?

The three boys went to the "show". The balcony that has been described was divided into two sections. One was accessible from the alley and the other by the stairs from the inner lobby. That section was known to be occupied by persons who were more interested in each other than the movie and that was the balcony section the boys entered. The other section was for the seating of members of the "Minority" race. They then climbed the partition dividing the balcony sections and leaned over the railing of the "Minority's" section while holding cans of vegetable soup that they had opened and began to call for their two friends, "H-U-U-U-U-GHY, B-I-I-I-LL, and poured the soup over the railing into the auditorium below.

The result is best described as a demonstration of mass hysteria.

Patrons seated directly below the source of the descending liquid with lumps in it quickly stood and proceeded to evacuate their section and seats in considerable haste which attracted the attention of other patrons who had no idea what the first group was fleeing, but never wishing to disagree with the majority, joined in flight from whatever was threatening the first group of people to scream and flee. The picture show patrons were escaping through the main entrance, the side door, and all of the emergency doors, so not knowing what the emergency was, the ticket agent placed a telephone call to the Evergreen Volunteer Fire Department.

When he received the call from the picture show the attendant at the City Hall, which was also the fire station, turned on the siren mounted on a tower beside the building, which was the audio signal to the volunteer firemen commanding them to report to the City Hall. Of course, everyone in town heard the siren and since observing a fire was considered to be spectacular entertainment the residents would lift their telephone receivers and wait for the operator to ask, "Number please?" The residents attempting to make a call would then ask, "Dorothy, where's the fire?"

Dorothy, the telephone operator replied to every caller, "I think it's the picture show."

After courteously thanking Dorothy and advising the rest of the occupants of the house of the assumed reason for the fire alarm going off, many residents got in their cars and drove downtown to watch. There were very few television sets in town to distract people at the time, probably three.

Of course, the former taxi drivers who were serving as policemen went to the scene to watch and to control vehicle traffic and the gathering crowd of spectators while "Pee Wee", Herston, and Ray remained out of sight in the theater.

When it became quiet the three boys sauntered out of the theater attempting to appear innocent, but there were too many people watching for them to succeed with their act.

When asked about it, "Pee Wee" will frown, look the inquisitor straight in the eye and claim, "This is the first I've ever heard of such a thing. Where'd you hear it?"

Ray and Herston have never expressed an opinion on the event and in 1969, a man walked on the moon and the dial telephone system was installed in Evergreen, Alabama.

There were times when he and Jan Hendrix would attend football or basketball games together, but Jewel, Mac, and Jimmy never went to one. Jan's father, Dr. R. W. Hendrix would sometimes drive by the McKenzie's home on Pecan Street and pick David up to give he and Jan a ride to the football field or gymnasium depending on whether it was a football or basketball game, not that it made any difference, both of them being at the High School. Neither Mrs. nor Dr. Hendrix was ever seen at one of the games anyway.

David's academic record had remained acceptable after the "encouragement" he had received early in the preceding year and he was initiated into the National Beta Club, a nation wide high school academic honor society and he continued to work part time in Mac's Repair Shop.

PART XIII

CREATIVITY SURFACES

Mr. Dey was teaching Chemistry, English and Literature, and American History where his lectures were completely and totally enjoyed, not only by David, but probably the rest of the class to. During history lectures he'd begin to play the part of characters in the stories and even invent probable dialogue while enacting two parts simultaneously by moving from right to left and back to right while assuming the different participants' roles. History was fun in Mr. Dey's class. He may have really been a frustrated thespian.

Jimmy's '23 Model "T" was designed and manufactured with the body having a wooden framing with the sheet metal body panels nailed to it and the entire assembly attached to the chassis by six bolts and since it had sat beneath the oak tree in Mr. Shell's yard for twenty seven years with no attempt being expended to protect it from the ravages of time and weather the wooden framing for the body had decayed and had no structural integrity remaining.

The body's wooden framing failed during the winter of '53 to '54 and in January 1954 David disassembled the body, piece by piece, but he did it carefully and kept and organized all of the loose parts. He then mounted an old boat seat on top of the fuel tank and the chassis could then be driven, but it certainly wasn't suitable for use as he searched the scrap pile behind the shop for pieces of wood that he could use to make new framing members. The parts that he made were an improvement over the rotted and failed parts, but their quality left a lot to be desired.

Mac admitted later that he never expected to see the car on the road again.

On the third Sunday in February David was extremely restless. He wanted to get out of the house and he didn't want to wait until 1:30 that afternoon to walk downtown to, "go to the show" that opened at 2:00 o'clock and he couldn't recall having driven an automobile since he had taken the body off the Model "T" the month before, so he decided to tell Jewel and Mac that he wanted to go to church. He was desperate.

He left home at 10:30 that morning in the '49 Ford Tudor sedan with very explicit instructions from Jewel to, *Go straight to that church. Don't dare go anywhere else. After church, come straight home.*

As he departed he thought, *Didn't she know that you can't go straight anywhere? Streets have curves and corners. Go anywhere else in Evergreen, and if you did someone would tell.*

He resented her demands and was slightly distressed about the lack of confidence and trust displayed by his mother and accentuated by her admonitions before he was given the keys to the car, but he had the car.

In the Presbyterian Church it didn't matter where he sat, Lizabeth McMillan, Mr. Mac's old maid sister who lived with him and who had been David's fifth grade teacher would always be behind him singing hymns in a very high pitched tone of voice using a volume that was not surpassed by anyone else. She was behind him again.

Dr. James Gailey, his wife, and their youngster had left the church and moved to Atlanta. Reverend Miller who was also a schoolteacher and at a later date taught at Evergreen High School and held his position in the church simultaneously replaced Dr. Gailey. Mrs. Miller was also a teacher and taught in Evergreen. David didn't remember anything Reverend Miller said in his sermon that day except something about Jesus being a bootlegger who was known for making wine that pleased the multitudes.

The services eventually ended and following Jewel's instructions to the letter, so far he went straight to the car, got in, started the engine, and headed for home, but not "straight home". He drove east on Carey Street, turned south on South Main, and finally right onto the Old

Knoxville Road which was a winding gravel road that ended at Bruner Avenue. He knew the road well and enjoyed driving along it because of the many curves and he could visualize it as a road racing course or hill climb course for racing cars.

He was driving fast, too fast, but he was being cautious for "blind" curves that he couldn't see around or across because of the potential for a collision with any other vehicles on the road when he arrived at a pair of curves that he pictured as a "chicane", an "S" turn, one to the right followed by one to left with no straight stretch between them. He entered the turn to the right and the car was drifting slightly as he looked across the curve he was approaching and attempted to make the change from a drifting right turn to a drifting left turn.

He turned the steering wheel more to the left and reduced the pressure he was applying through his right foot to the throttle pedal and the rear of the car started to swing outwards to the right. He started to rotate the steering wheel back to the right and to add some pressure with his foot on the throttle when he realized he no longer had control of the car. His "survival instinct" took command and he moved his right foot to the brake pedal and pressed HARD.

The rear of the car continued its rotation around the front towards the right until the vehicle was sliding sideways along the road and gradually approaching the ditch that paralleled the road. The rotation ceased and the front wheels dropped into the ditch, crossed it, and the front end smashed into the embankment on the opposite side of the ditch from the roadway, climbed the bank, and collided with a tree at the top of the bank. The rear of the car then swung to its right and into the ditch which caused the car to roll onto its right side and it continued to slide along the ditch going rearwards on its right side with the roof against the bank until it reached a point where the ditch deviated away from the road and the right end of the rear bumper wedged in the corner of the ditch causing the car to stand on its rear end. It stopped momentarily standing on its rear end and then fell back into the road on its wheels, upright and stopped. It had spun two hundred seventy degrees, crossed a ditch, climbed a bank, struck a tree, skidded backwards on its right side, and tumbled to an upright position sitting crosswise in the road.

David looked around, noticed a pain in his left arm caused by his impact into the left door when the car stopped and then thought, "Maybe it's only scratched up and I can polish it out." Even as the thought ran thought his mind he realized how absurd that hope was. The door wouldn't open when he pulled the release handle, so he had to rotate in the seat, place his feet against the right door, his back to the left door, pull the handle, and force it open. When he observed the door sag below the rocker panel, which is the panel that forms the lower surface of the door opening, he knew that the chassis frame was bent. He got out and walked around the car inspecting it and could see that the only body panels that weren't damaged were the left door, left quarter panel, and deck lid. The car was a total wreck.

He walked from the accident site to Bruner Avenue where he turned north and walked to his classmate, Herman Coburn's home, gained admittance and access to the telephone, answered the operator's request saying, "258-J please, ma'am." Five parties that included both Snowden families shared the telephone line and number and it was *busy*. *Probably some jock strapper pursuing one of the Snowden girls*. He thought.

It was nearly 1:00 o'clock on Sunday afternoon and he was sure his parents would be concerned about the car and him, so: after several attempts to telephone his parents and finding the "number busy" each time, he asked the operator to interrupt the parties using the phone and allow him to speak to his parents. "This is an emergency," he said.

Mac asked if anyone was injured, and then assured him that he'd drive the Model "T" chassis to the Coburn's home and pick him up. It assuredly presented a memorable spectacle in the streets of Evergreen. A Model "T" chassis with the only sheet metal in the entire collection of parts being two front fenders, two running boards with their splash shields, a firewall, and a hood with two males, one middle aged, and one a teenager, seated on the fuel tank that was normally under the front seat, and both of them exhibiting very stern, determined, and serious expressions on their countenances.

They drove to the accident site where Mac inspected the wrecked sedan and then went to the house on Pecan Street which David was mentally bracing himself for entering.

Jewel asked, "Are you hurt?"

"No ma'am."

"Give me that driver's license."

David removed the license from his wallet, but he never had an opportunity to surrender the document. Jewel ripped it from his hands, tore it into pieces, and hurled those into the trash can located under the kitchen sink, slammed the cabinet door, and stood in front of him again and shouted, "Are you sure you're not hurt?"

"I'm not hurt." he said meekly.

She reached out with both hands, jerked his glasses off, and slammed them down on the kitchen table. He thought she probably broke them.

"You're sure you're not hurt?" she enquired again.

"I'm O.K."

Jewel had restrained herself as long as she possibly could. Her explosion was immediately in front of the boy and her hands were moving so fast that they appeared as two blurs, but they did stop alternately on his shoulders, on his head, his back, and his chest. When he put his arms up to block her blows she initiated the use of a vocal attack.

"Get those arms out of my way. I'm gonna' kill you." It was good that she used her bare hands instead of a weapon like a paddle or a belt. If she hadn't she probably would have fulfilled her promise and threat to kill him. She was much rougher than the wreck itself.

She eventually wore herself down, physically and emotionally. He was surprised to realize that he was still standing and while all of that was happening Mac had been talking on the telephone to Byron Warren, the manager of the local Ford dealership that operated the only tow truck in town.

Mac said, "Change your clothes and open the shop. The car will be here after Byron Warren finishes his Sunday dinner." When they closed the shop at 6:00 o'clock that evening the car's frame had been straightened, a new radiator was installed, the cooling fan straightened, and contradictory to its appearance, it was drivable.

As usual, Mac dropped the boys off at school the next morning and when David arrived home that Monday afternoon he was told that his weekly salary was being reduced from five dollars to fifty cents per week until the parts required to repair the car were paid for.

Even in 1954, fifty cents per week was inadequate for a seventeen-year old high school student to support any social life.

And so, by May of 1954 he had completed constructing the replacement wooden framing for the Model "T" body, reassembled it, had his driver's license returned by his mother, and was trying to find a way to fuel the Model "T". The price of gasoline had risen to twenty-one cents per gallon and the Model "T" only delivered approximately twenty-two miles per gallon of fuel.

As has been stated, Mr. Dey ran the Chemistry class at the rate determined by the student's rate of absorption and level of comprehension. By the middle of April 1954 the 9:00 o'clock chemistry class had completed the textbook and needed something simultaneously educational, challenging, and creative to distract the student's thoughts during the class period.

Mr. Dey told the class to break up into small groups and select a subject of study and experimentation, submit the proposals for his review, and hopefully, approval. Jan and David already close friends for the past eight years, teamed up and due to David's need for an "economical alternate fuel" for the Model "T" they selected, "Alcohols and Ethers" as a potential subject for study.

David had no idea that he'd be involved in an attempted solution to the same problem at Ford Motor Company in 1973 through 1975 while employed in a department named, "Alternate Engines Research."

Mr. Dey approved the project.

Their first challenge was to identify something readily and economically available to ferment which was also a locally plentiful substance. Molasses made from sugar cane syrup met the requirements. Maple syrup was costly and in limited supply and probably wouldn't support fermentation as well.

The second stage in their strategy was to dilute the molasses with plain tap water at a ratio of four parts water to one part by volume of molasses and then add three small yeast cakes per gallon of the mixture which they placed in a one gallon clear glass bottle and devised a pressure relief valve of glass tubing and water, placed the apparatus in the window and waited.

In approximately a week and a half the bubbles stopped appearing in the water in the pressure relief valve and they knew that fermentation was completed. They mixed lime filings into the fermented liquid and then filtered the fluid through a paper filter and were ready for the next step. At that point the filtrate fluid was showing readings indicating that it was forty-proof (twenty percent) alcohol by checking with a hydrometer.

During the third week of the experiment the partners set up a boiler that exhausted its vapors into a condenser and ignited the Bunsen burner beneath the boiler. They soon saw droplets of the product coming from the condenser and the other students in the class who had been complaining about the plumbing being attached to the only faucet in the room became fascinated and even enthused about the project. There was also a sweet, gentle, and very pleasant odor wafting through the High School end of the building that everyone who sensed it seemed to appreciate and enjoy. Yes, it was a heavenly aroma better than flowers in the spring and far superior to perfume on old ladies in church. It did not reach the Junior High School end of the facility. The hydrometer test indicated that the product was 180 to 190 proof alcohol.

On the morning of the third day of operation of the still Coach Parsons, the replacement for Coach Law who later married one of the Snowden girls came into the room. He immediately turned around and departed. He did not belong in the chemistry class anyhow, but he came back accompanied by Coach Hart and both of them left the room. The two, "jock-strappers" returned accompanied by the school principal, Mr. Claybrook, who then invited Mr. Dey to accompany all of them to his office.

Jan and David were ordered to disassemble their refinery for an alternate automotive fuel and to select another subject for their studies and research which led them to believe that naphtha gas might be a potential alternative fuel.

At 9:15 on the morning of the first Friday in May 1954 the budding researchers for an alternative fuel for internal combustion engines ignited the Bunsen burner beneath a beaker of naphthalene mothballs hoping to liberate naphtha gas from its confinement in the mothballs and they had decided that appropriate disposal of the freed gas could be achieved by igniting it and allowing it to burn freely at the end of the glass nozzle they had made and installed on the top of the beaker.

In Evergreen High School during the early 1950's it was accepted practice for the co-eds to appear in class on Friday mornings with preparations for their evening dates in an advanced state of development. Each of them would be dressed in a white shirt that they obviously borrowed from their father's wardrobe with the sleeves rolled halfway to their elbows, the top two or three buttons below the collar left open, and the tail hanging loose around the waist, not tucked into the blue jeans they wore with the legs rolled halfway between the ankle and the knee, socks, tennis shoes, and the entire costume topped by their hair in rollers. They intended to LOOK GOOD on their dates with the "baby jock-strappers" that evening.

In later years the corporate world would adopt a practice in its offices called, "casual day" that reminded Dave of those Fridays. The girls displayed noticeable similarities and characteristics in attire with High School girls from the mid 1950's.

The mothballs melted in the beaker and then began to boil releasing the naphtha gas in spurts through the nozzle into the atmosphere. The flow of the freed gas soon became smooth and steady and one of the researchers ignited the escaping vapor.

It was impressive. The free flame was approximately twelve inches long and three or four inches in diameter, but there had been no attempt made to control the mixture for the ratio of oxygen to flammable gas in the experiment and there were strings of soot rising towards the ceiling and then rolling along the ceiling to the walls of the room where they changed direction following the walls to the corners where they accumulated and then descended in the corners as the air supporting them cooled.

The students in the class except for the two researchers fled to the corners of the room prior to the strings of soot arriving overhead in the same location. They were attempting to retreat as far as possible from the site of the spectacular experiment and the corners were located the greatest distance from the source of the soot.

And then the strings of soot started rolling down the walls and of course, the students were already there and they were complaining so loudly that they couldn't hear anyone's recommendation that they get out of the corners. Most of the noise was being generated by the co-eds who didn't want the strings of soot in the hair that they had so carefully washed and rolled in preparation for their dates with the, "baby jockstrappers" scheduled for that evening.

Coach Wendell Hart, Coach Bill Parsons, and Mr. Clinton Claybrook soon appeared in the room and all of the students except the two researchers, Jan and David were excused from the class for the day. The researchers were directed to dismantle the experiment, clean the equipment, and report to the Principal's Office.

The fact that Jan's father, Dr. R. W. Hendrix was the County Health Officer and David's father, "Mac" McKenzie was a respected member of the community undoubtedly influenced the school administration in its selection of a response to the situation they believed was potentially hazardous to their reputation. Mr. Dey advised Jan and David that they were to report to the Principal's office at 9:00 o'clock every morning for the remainder of the school year which had its usual routine of final exams and relaxed demands in classes.

Mr. Dey encouraged reading in his English class and gave credit to the students who browsed the library at school and the Evergreen Public Library for books to read and then give verbal reports to the class on each one they read.

There wasn't much in the Public Library which had been a gas station on McMillan Street at the intersection of U.S. Highway 31 and the L & N Railroad until the Works Progress Administration (WPA) had built the railroad overpass known in the town as, "the bridge" and effectively isolated the small building which some people said had also served as the Greyhound Bus Station in its heyday.

A group of ladies in the little town had gotten together and formed a group they called, The Evergreen Study Club, acquired the use of the abandoned gas and bus station, and revamped it into Evergreen's version of a library. Those ladies deserved a lot more recognition and appreciation than they ever received for their effort, foresight, and civic pride that created a community asset without the so-called assistance of "bureaurats" and the taxpayer's financial commitment.

Near the end of the school year Mr. Dey had assigned a requirement for a term paper in his English class and allowed the students to select the subject of the composition that each would offer. David elected to present a biography of the early exhibition pilot, Lincoln Beachey for his subject. In the paper he mentioned another pilot, Bob Fowler.

Mr. Dey collected the papers and sometime during the next week he advised David that he had met Bob Fowler when he landed near Evergreen while he was making his trans-continental flight from the Pacific to the Atlantic Oceans in 1912. Mr. Dey told David that he was a young boy when Bob Fowler landed near Evergreen and was taken to the landing site by his uncle. He suggested that he and David visit the Probate Judge's Office where back issues of the local newspaper, The Evergreen Courant were on file in the vault and perform a search for the article that was published.

In order to take the afternoon off from his part-time employment by his father David had to explain in great detail what the objective of the visit to the Judge's Office concerned. Jewel and Mac had not forgotten the wreck of the '49 Ford sedan. The search was successful and David never forgot it.

What had been a busy school year wound down to summer vacation for most of the students, but David had to work in, Mac's Repair Shop.

During the '54 to '55 school term it was David's senior year and his social activities were only attending the football and basketball games. He preferred that to remaining at home.

His class chose to make the, Senior Trip to New Orleans instead of the traditional trip to our nation's capitol and needed to raise funds to support it. David's was not the only financially cautious family in town. The class was allowed to operate the concession booth in the gymnasium during basketball games and retain the profits from the soft drinks, popcorn, and packaged treats that were sold to the fans to financially support the Senior Trip.

Hygiene was not a primary concern in the concession booth. Every product sold except popcorn was pre-packaged, therefore no special effort was considered essential to assure cleanliness. After all, no one spent a great deal of effort on that endeavor at football games. That management oversight almost bit them one evening in the gymnasium when the popcorn machine refused to operate. Because David worked in his father's shop and Mac McKenzie seemed to be the master of anything and everything then surely David inherited the same talent or so people in Conecuh County thought. He was immediately called upon to diagnose the problem with the recalcitrant popcorn machine and perform remedial procedures to achieve remarkable results inspiring the unit to resume productive and profitable functions.

He found that the machine did not appear to have ever been washed or cleaned. It appeared that all of the insulation and connectors on the electrical wiring were soaked in grease or oil used for preparing popcorn, so he cleaned up all of the plugs, receptacles, and as much of the wiring as he could reach and after about fifteen minutes of effort during which time a line of potential purchasers had formed outside the booth he commanded, "stand clear" and turned it on.

There was a loud noise like, "PFFFFFT", sparks flew, and the lights in the gymnasium went out. That was followed by a great, "WOOOOOOOO!" from the crowd of occupants of the gymnasium and then the lights came back on. The popcorn machine was running again and everyone was staring in his direction. He smiled and made a "palms up" gesture with both hands towards the machine as though to say, "There you go." He never received a, "Thank you" or any other rewards for his efforts.

There was nothing that he believed was spectacular or memorable that happened in David's senior year of High School and he didn't recognize the wonderful opportunities that did present themselves that spring. He had no hope of going to college and really had no idea what he would pursue after graduation. He had wanted to drive racing cars, but he had no car, no experience, no sponsor, no reputation, and no obvious means of acquiring any of those qualifications. He may have wished to design automobiles, but that obviously required some college training and he had no hope of attending college. He didn't have the financial resources to support that goal and had none in sight or mind. He had abandoned any hope of doing anything in aviation because there was no encouragement from anyone for that objective.

In April 1955 he was allowed to visit the college campus at Auburn with Glenn Pate, Don Pate, Carness Persons, and John Wilson to attend an "Open House" intended to provide high school seniors an opportunity to see a college campus and hopefully encourage them to make arrangements to attend college at Auburn. While there he visited the Engineering Extension Service office and learned of the Co-op" program which allowed a student to attend college one quarter and then hold a job in a related industry on alternate quarters and hopefully, by carefully managing his funds earned thereby, acquire a college degree in six years instead of the normally anticipated four years. He applied for admission to Alabama Polytechnic Institute, now called Auburn University for the summer quarter of 1955.

Soon after that visit to Auburn Jewel and Mac sat down with him and explained that they had borrowed three hundred dollars from the Conecuh County Bank and that loan combined with five dollars per week that they'd mail to him weekly would be their contribution to his college education. He was also told that, "If you can't carry it by the end of August that'll be the end of it."

During the third week of May the senior class made the traditional Senior Trip to New Orleans instead of the usual tour of "Malfunction Junction" (Washington, D.C.). Coach Hart was assigned as one of the Faculty Chaperones, a job that David didn't believe he performed very well. David did enjoy the trip and visited the city many times later in his life.

On May 31, 1955 he graduated from Evergreen High School.

PART XIV

COLLEGE YEARS BEGIN

On June 12, 1955 Mac drove Dave to Auburn in the '49 Ford, helped him take two trunks of clothes, bed linens, and miscellaneous other items believed to be needed in the dormitory room from the parking lot to the door, shook hands, and drove away. The next day he began the enrollment process at Auburn by beginning three days of scholastic placement examinations which he did well on, registration, and class scheduling.

On Thursday at 7:00 o'clock in the morning he began his first class in college. It was the course in Plane and Solid Geometry that he was required to take because Solid Geometry was not in the curriculum at Evergreen High School. That was followed at 8:00 with College Algebra. Then he had to walk a half-mile to the next class at 9:00 o'clock, which was English Composition for Advanced Students. He then had to walk another half mile to attend classes in "Air Science" which was the classroom part of the Air Force R.O.T.C. program Taught by an Air Force officer, Colonel Finch who had been a P-38 pilot in North Africa during WW-II. David liked Col. Finch very much probably because Col. Finch seemed to like him.

After going back to the dormitory cafeteria for lunch David then had to go to Ramsey Hall for Engineering drawing classes. It seemed to him that the students taught them selves by reading the book, conferring with each other, and only accepted the assignment from the Instructor, Leon Marr Sahag. Mr. Sahag was an immigrant and had such an accent that when combined with their own linguistic backgrounds made comprehension of his lectures completely impossible.

One afternoon per week he had a three-hour long class in the Foundry Lab. It was one of the Engineering Shop courses and presented the concepts of producing metal parts by pouring molten metal into a sand mold of the desired part that completely captivated David's interest. Of course he already knew something about sand casting, but this was his first close exposure to the process.

All of the courses added up to a total of twenty quarter-hours for the summer quarter of 1955 and was David's first quarter in college. It was quite a load and an ambitious endeavor for a First Quarter Freshman. Of course he had immediately met with the Director of the Engineering Extension Office and enrolled in the Engineering Cooperative Education Program.

Within two weeks of beginning his first quarter at Auburn he responded to a note he read on a bulletin board announcing a meeting of the Auburn Spots Car Club in one of the rooms in Biggen Hall. He attended the meeting and felt like a child who should never have left his playpen, but he decided to be quiet and attempt to bluff his way along. He didn't realize that such a tactic wasn't necessary and his problem was his own bashfulness and self-consciousness. The psychological problem was most likely a lack of confidence and it never occurred to him that it was the beginning of a penalty he was about to pay for immaturity. It was the beginning of a social life he didn't realize should not have been selected. The social life that he was becoming involved in wasn't illegal, immoral, dangerous, or unacceptable. It was only something that would not have been recommended by a counselor at that time to anyone with his immaturity and inexperience.

He was very conscientious about attendance and punctuality and didn't participate in the activities at the two bars or nightclubs located on Highway 29 south of town, but this was his first exposure to a life free of close supervision and he was still distracted from his studies by modified automobiles and their owners. There was also a half-mile long oval shaped dirt racetrack located halfway between Auburn and Opelika, Alabama that was owned by the Garret family. It didn't take him long to "sweet talk" Mr. Garret into letting him go into the pits on Sunday afternoons during the stock car races. He wanted to drive one, but fortunately he was too shy to attempt asking car owners for a "ride". He knew that it would lead to frustration, disappointment, and be non-productive to ask for a "ride". He had no qualifications that would justify the owner of a car risking equipment in his hands and he knew it. He knew it was undoubtedly beneficial to his mortality continuation and overall physical condition to be left out of stock cars at that time.

During a meeting of the Auburn Sports Car Club in July of 1955 one of the members, Ed Black organized a Scavenger Hunt" for the evening's program and David, since he did not own an automobile of any type rode in a '53 Corvette with its owner, Bill Freel. It turned out to be the high point of the summer's recreational activity.

After picking up the list of "treasures" to seek he and Bill had no difficulty driving north of town and reading a particular road sign or obtaining a ticket from the Auburn-Opelika drive-in movie, but obtaining a pair of ladies panties presented a genuine challenge.

The passage of time has obliterated any possibility of accurately recalling who suggested going to the women's dormitories and asking for a pair, but it befell David's responsibility to enter a dormitory and make the request while Bill remained in the car with the engine running at the curb awaiting his return or the development of other responses. David had no idea what he would have done if the response were anything other than approval for the request and suspected that Bill would have fled without him in the car if there had been any hint that such action was appropriate. He was pleased and relieved to find Bill awaiting his return when the desired objective was acquired.

It's still difficult to believe that he did such a thing. He was eighteen years old, shy, bashful, and had only two dates with a young lady while in High School. Perhaps the threat of failing to acquire the pair of "unmentionables" and the accompanying revelation of his shyness and immaturity was potentially more embarrassing than having to face a homo-sapien of the feminine gender and make the request.

Upon returning to the building, Biggen Hall and room where the meeting was being held they were pleased to see that no other team had returned with as many "treasures". The other members of the ASCC proceeded to reveal their "loot" for counting and evaluation and when the tally was being compiled there was a lot of laughter and speculation exhibited about how the team of Freel-McKenzie had acquired one item in the "treasure" hunt. No other team had acquired a pair of "unmentionables" intended for feminine wear.

They had won, but there were neither prizes nor declaration of a winner. There was no recognition. He had forced himself to present the requirement to a young lady at the reception desk in the dormitory and then suffer the same trauma a second time in the presence of the House Mother which led to the acquisition of the required item without experiencing arrest or verbal assault. Profit for the assumption of those risks was non-existent. That wasn't the only time he was not given the recognition and compensation that he felt was deserved and earned. Honestly, it became a regular and frequent occurrence in his opinion. His interest and ambition in auto racing combined with his immaturity to interfere with his academic achievements.

The Engineering Extension Service Office had arranged interviews with Chemstrand Corporation, The U.S. Government, and Kimberly-Clark Corporation seeking employment as a "co-op student", which led to written examinations being administered and graded by Chemstrand and the U.S. Government. Kimberly-Clark had two of the applicants appear in their offices in Memphis, Tennessee for further interviews. One of those applicants was David.

The Chemstrand Corporation chose to withhold any offer for employment and ultimately, he refused an offer for employment at the George C. Marshall Space Flight Center, Redstone Arsenal in Huntsville, Alabama. At the end of his first quarter in college he was flown to Oshkosh, Wisconsin and then traveled by automobile to Neenah, Wisconsin and interviewed and hired by Kimberley-Clark Corporation to work as a "co-op student" in their Memphis facility for the fall quarter, September through December 1955.

The series of flights from Montgomery, Alabama to Nashville, Tennessee, to Louisville, Kentucky, to Indianapolis, Indiana, to Chicago, Illinois via Eastern Airlines, and then via Wisconsin Air in a DC-3 to Milwaukee and Oshkosh, Wisconsin was the greatest distance he had ever traveled and was uneventful. He was met at the airport in Oshkosh, later named Wittman International Airport by the Chief Engineer of Kimberly-Clark Corporation and his wife then driven to a hotel in Neenah, Wisconsin.

The next morning he was picked up by a company car and spent the first half of the day in interviews, on tour of some of the company facilities, taken to lunch, and hired. He was then transported back to the airport in Oshkosh where he had to wait at least an hour for the flight to Chicago.

He always had an aversion to waiting and when he looked out of the doorway of the very small, one room "terminal building" he could see a hangar that was open, but couldn't see anyone in the hangar as he walked towards the irresistible attraction. He had been around "grown up's" toys for all of his short life, eighteen years, so he knew better than to touch anything and decided to gamble on presenting his credentials if anyone challenged his reason or right to enter the hangar.

Inside the hangar he found wings, engines, fuselages, and what appeared to be complete airplanes hanging on the walls, from the roof trusses, and scattered about on the floor. Somewhere in the collection of components he saw the name, "Wittman" painted on something and remembered having read that name before, somewhere, sometime, and knew that he was in the hangar of a celebrity in air racing.

In the center of the hangar sat a small, red, high wing monoplane that obviously wasn't an old airplane in spite of it being apparently quickly built without much time, money, or effort having been spent on paint and finish. He leaned over to look in the left side window while exercising caution to avoid touching the little airplane and noticed that there were three people in the office attached to the hangar and one of them repeatedly came to the personnel door to the hangar and looked in his direction.

He decided to hurry his walk around the little airplane and leave the hangar so that he wouldn't be, "yelled at".

He boarded the Wisconsin Air DC-3 and was flown to Chicago where he then boarded a twin engine Martin called a, "Silver Falcon" by its operator, Eastern Airlines and departed enroute for Atlanta, Georgia with stops scheduled in Indianapolis, Louisville, and Knoxville.

When approaching Indianapolis the pilot had to reduce the power settings on the engines to descend and at some point he also had to lower the nose of the airplane tilting the aircraft forward. The passengers only noticed the smoke that started entering the cabin from the heater and air conditioning ducts, not the change in the aircraft's attitude. Everyone looked towards the stewardess, they're called "Flight Attendants" now and felt reassured and confident of their safety when she signaled those who were giving her inquiring looks that everything was, O.K..

There were a few passengers that deplaned and a few more that boarded in Indianapolis. The Stewardess explained to a passenger that inquired about the smoke that it was the result of an oil reservoir in the air conditioning system being overfilled and when the excess oil was depleted it would stop smoking. David relaxed and simply sat in his seat while the airplane took-off, flew to Louisville, began the approach, and the identical series of events occurred including the smoke coming from the vents with the same response from the passengers that boarded in Indianapolis. The same events were repeated in Knoxville and Atlanta where he was driven downtown to Atlanta's premier hotel.

The next morning he arrived at the airport at the last possible moment to dash across the parking ramp and board the Martin, "Silver Falcon" and relax into his seat on the flight departing for Huntsville, Alabama, Birmingham, and Montgomery where he would deplane. Unlike the day before, the airplane was fully loaded and had no empty seats.

While he was growing up Jewel had always demanded that he wear a sport coat and a tie whenever he left Evergreen and when he was interviewed at Kimberly-Clark that dress code requirement was accepted and he was not comfortable if he violated it. The collection of passengers on the flight from Atlanta to New Orleans with stops in Huntsville, Birmingham, Montgomery, and Mobile were dressed absolutely anyway they were in the mood to dress, but there were two Nuns, and three merchant seamen who became especially noticeable for other reasons, too.

The two Nuns were wearing their Habits and their deportment was exactly what one would expect. The three merchant seamen were wearing dirty shirts and blue jeans that may have never been laundered since they purchased and started wearing them. They had originally departed on their trip from New Orleans and their final port had been Boston, Massachusetts where their employer had purchased their airline tickets to return to the port of origin, New Orleans. They had stopped overnight in Atlanta and had spent the evening, "bar hopping" which explained their appearance aboard the airplane while still as drunk as boiled owls. That information was all made available by listening to their remarks, which was impossible to avoid.

David wondered why they were allowed aboard the airliner while either intoxicated or suffering the discomfort of recovering from that condition. The instant the airplane reached its cruising altitude and the "fasten seat belt" light went out the sailors began another party and soon discovered the Nuns.

The three merchant seamen gathered at the Nuns seats and immediately proceeded to attempt to, "de-flower" them and their remarks soon degenerated into inappropriate, unacceptable, and even obscene suggestions. The scene was embarrassing to David and even though he was not one to get involved in confrontational situations he was beginning to consider it, but the consequences that had to be considered precluded the likelihood of there being much benefit to anyone by challenging the merchant seamen. He was really getting nervous when the airplane's nose lowered and smoke began to enter the cabin through the vents.

David thought, "Hooray, I know this airplane." as the three merchant seamen immediately dropped to their knees, folded their hands, began to beg, began genuflecting, and suddenly became sweet children.

David was probably not the only passenger that was amused, and releaved. He deplaned in Montgomery, caught a taxi to the Greyhound Bus terminal, and returned to Evergreen, "via Greyhound".

One week later Mac purchased a bus ticket to Memphis, "via Greyhound", gave David twenty dollars, shook hands, and the next stop in seeking a college education began.

Jewel's brother Gilbert Currence who the family members called, "Mark" lived in the Ambassador Hotel on South Main Street in downtown Memphis which David selected as his first destination and having visited his uncle during the interview with Kimberly-Clark in July, knew he could easily walk from the bus terminal on Beale Street to the hotel. He and his uncle reviewed the advertisements in the local newspaper to find, "room and board" then made a few telephone calls in response to the advertisements, selected a likely room, and rode to the address in a taxi. His uncle had never learned to drive an automobile.

David decided to rent the room in the first home they inspected, made arrangements with the landlady for credit until he received a paycheck, and moved in.

The next morning, Monday he walked to the nearest bus stop and rode the City's Transit Bus system to the Kimberly-Clark Co. on Thomas Street in the northwest side of town and began his first "co-op" job assignment. He was acquainted with his aunt and uncle, had just met his new landlady, her son who was a couple of years older than he, and a few employees at the mill by the end of the day on Monday. He was a stranger alone, financially restricted, had no transportation, and thought his future was completely open for opportunities. There was certainly nothing on his agenda to limit his search for opportunity, challenge, and entertainment.

He'd work eight hours per day, five days per week, and he'd sleep very late on Saturday and Sunday mornings, go to a movie on Saturday evenings, visit his uncle on every third Sunday, and looked forward to Monday mornings and resumption of the work schedule.

His landlady and landlord's son, Charles, a high school drop out two years older than him, attempted to include David in his social life, which was to hang out at the nearest billiard parlor and a corner gas station. There was nothing that seemed to be illegal or dangerous about it, but there was nothing of interest to David at those sites. He did discover the local hot rod club, The Memphis Rodders. He didn't own a car, didn't have the resources to obtain one, and all of the hot rodders knew it, but they allowed him to join and attend the meetings every Sunday evening on the second floor of a small restaurant on Sumner Avenue.

After two months of living in the home on Forrest Avenue not far from the, Crosstown Theater he decided that Charles, the landlady's son was probably going to cause some type of domestic trouble, so rather than be involved he moved to another boarding house on North Parkway Boulevard where he had a roommate, Benny Massey. after two more months in Memphis where he had managed his finances well he left financially prepared for his next quarter at Auburn.

Once again he was riding on, Greyhound and arrived in Evergreen on December 24, 1955 where he spent Christmas with his family and on January 2, 1956 he was on the campus at Auburn again registering for another heavy load of classes. He enjoyed all of his classes except Chemistry and Physical Education. Once again he was residing in the men's dormitory that he referred to as, "Mongolian Hole". Magnolia Hall was not an undesirable place to reside, but students will always exercise the liberty to contribute their personal nomenclature to anything. It seemed to present too large a social distraction for a student, though. For entertainment he made the mistake of resuming his prior association with the Auburn Sports Car Club.

No one remembers who made the proposal or initiated the expenditure of the time, effort, expense, and thought in the group of students in the ASCC and contacted the farmer who owned the property located North of Tuskegee which was a cow pasture that contained the runways that were the remains of Sharpe's Field, the home or training base of the Tuskegee Airmen during World War II, but the farmer allowed the ASCC to sponsor a drag race at the site.

In February 1956 the news had spread and on a Sunday morning there appeared at Sharpe's Field lines of cars filled with spectators and many competitors. David had ridden to the site with Perry Dillman's stepson, the owner of a new '56 Ford who was interested in drag racing the car. Perry had been a flight instructor at Sharpe's Field during World War II. Some of the ASCC members stood at the entrance to the pasture and collected admission fees while others measured the quarter mile section of runway to be used as the drag strip and marked the return lane, designated a pit area, and generally organized the event that functioned remarkably well except for a problem with spectator control.

There were no grandstands, barriers, guardrails, policing, or any provisions at all made for spectator control or safety. They simply stood on the abandoned airport runway beside the area that the cars were running on. The spectator line began beside the starting line and as the day passed the spectator lines grew longer and longer. Of course the spectators who were the farthest from the starting line began to move forward in an effort to see the competitors and the flagman at the starting end of the strip. The spectator's drifting forward caused the cars to be accelerating into an area that grew narrower and narrower as the cars progressed along the designated drag strip and their speeds increased. It was a potential disaster begging to occur.

Periodically the ASCC members had to stop the drag races and four of them would drive down the runway between the spectators and attempt to force them back from the path the competing cars were following. The spectators' response was varied. Most of them backed away, but there were some who were angered by the interruption in the races and some who resented being asked to move back for their own safety. It was another demonstration of mob mentality. Those people had traveled to the site of the drag races to view them, had paid a dollar for that privilege, and now were going to see the action whether a car had the audacity to kill them or not. It was unbelievable.

During one of the passes to attempt forcing the spectator line back a few feet one of them pressed his pocketknife against the side of Bruce Lund's beautiful customized '55 Ford hardtop causing serious damage as he drove by. Of course the perpetrator retreated into the crowd and disappeared when Bruce stopped, got out of the car, and attempted to challenge him. It was a dramatic display of the behavior that gives a bad name to "red-necks and hillbillies".

Miracles do occur, occasionally. That's the only reason there were no injuries at the ASCC Drag Races on Sharpe's Field in February 1956.

The ASCC had no sanction, treasury, legal standing, authorization, or recognition. Legally, there was no such organization and if there had been an accident, responsibility would probably have fallen on the person with the least ability to escape. Fortunately, a miracle occurred. No one was killed or injured.

The members of the Auburn Sports Car Club met on the campus that evening after the races were concluded and divided the proceeds from the drag races which provided each member approximately a hundred ninety dollars.

Classes proceeded as usual for David and he worked on his studies sufficiently to respond to demands that he couldn't avoid and as might be expected, the scores did not reflect his real capability.

During a meeting of the ASCC a couple of weeks after the drag races one member, Joe Cooley announced that he held in his possession an eight millimeter black and white "smoker" movie and a projector. The meeting promptly adjourned as the attending members hurried to their vehicles and reconvened at Joe's room that he shared with five other students, none of whom were present. It was a converted attic in a home a few miles southwest of Auburn and had stairs on the exterior of the southern end of the house with the sole window at the northern end of the gable roof.

The ASCC members rapidly climbed the stairs, entered the room, arranged chairs, set up the projector, removed a sheet from one of Joe's roommate's bunk, and hung it over the window end of the room for use as a projection screen.

Never before or since has David seen such a movie. It was blunt, and what is now called, "hard core" pornographic. At the end of the second showing he and whomever he was riding with left the remainder of the group, the film, and the entertainment. The two of them were undoubtedly driven by embarrassment, shame, and self-consciousness to disassociate themselves from such activity. As they proceeded towards town they met three cars headed southwest. The first car was marked as a City of Auburn Police car. The second was marked as an Auburn Campus Police car and the third was marked as a Lee County Sheriff's car.

The next day he heard that they had forgotten to close the window shade before placing the bed sheet over the window to use as a projection screen and the neighbors had an equally excellent viewing of the movie. They and their children apparently were not as well entertained by the movie, so they called the police.

It was the last meeting of the Auburn Sports Car Club that David attended.

David had learned to drink coffee and to smoke cigarettes, which really aroused Mac's sense of revulsion when he learned of the smoking. When he was twenty-eight years old, David came to his senses and quit smoking after he realized that there was some truth in Dr. Hendrix's assertion that smoking tobacco was the American Indian's vengeance on the white man. Dr. Hendrix claimed that during the first few years of their exploration of the American continents the Europeans had given the Indians syphilis and Christianity, therefore the Indians repaid the white man with something of equal value, the addictive use of tobacco and its accompanying effects on health and life expectancy.

He hadn't drunk coffee at home or anywhere else in Evergreen. He simply never had the desire to do so. He had always noticed that adults habitually consumed significant amounts of the elixir and when he was approximately five years old he saw Jewel and some of her friends drinking coffee in the kitchen one morning and asked for a cup of coffee for himself.

Jewel responded, "Well, O.K., but if a growing boy drinks coffee, it'll cause him to be black when he grows up."

David lost interest at the time, but while in college he learned to accompany his acquaintances in the consumption of coffee.

At the end of the Winter Quarter, 1956 he didn't have a job on the "co-op" program, so returned to Evergreen and resumed working for Mac. That spring Mac built the shop at the corner of Belleville and Pecan Streets and David observed an example of vindictiveness that inspired a resolution that lasted for the rest of his life. "Don't hire the local attorney when confronting the local *bureaurats*."

Jewel and Mac had purchased the lot on the corner of Belleville and Pecan Streets in an area zoned, Light Industrial that was an appropriate area for occupancy by a welding and general repair shop. As usual Mac didn't make any drawings of the structure. He didn't have to. He had an absolutely amazing ability to create and store in his mind an amazing amount of information and structural detail of anything he wanted to build or repair.

David made a drawing showing the three basic orthographic views and a perspective view of the structure that was used with the application for a Building Permit. The Building Permit was issued and construction proceeded with Mac supervising the concrete and masonry workers that he hired and David kept Mac's Repair Shop open. Incidentally, the brick masons that Mac hired were of African ancestry.

After the masons had completed the cement block walls and other friends had brought in their equipment that specifically was a logging skidder and loader and lifted the "I" beams into place to support the roof rafters David was put into service helping to construct the roof. His day began at 7:00 o'clock in the morning and at 10:00 o'clock he left the construction site and opened Mac's Repair Shop. At 12:00 he ate dinner after which he returned to the shop where he remained until 5:00 o'clock in the afternoon and then he'd close the shop and return to the roof at the construction site remaining until 8:00 o'clock in the evening. It may have inspired him to give more serious thought to a college degree.

Mac got out of the bed and supervised, worked, and did a lot of the manual labor, himself. David was essentially an unskilled laborer working on the building and in Mac's Repair Shop.

While they were on the roof one morning the only police car owned by the City of Evergreen arrived and was being driven by the, Chief of Police whom Mac climbed down to converse with. The Police Chief was delivering a Stop Work Order and a notice that the Building Permit had been revoked. Mac and David did not stop working on the building and Mac proceeded to meet with the Mayor and each of the five City Councilmen individually which led to a climax at the next regularly scheduled meeting of the City Council.

The City's governmental officials asserted that they were concerned about the possibility and likelihood of a fire or an explosion from a welding shop within the corporate limits of the community. There was one other welding shop in Evergreen at the time, but its owner was not attempting to relocate and had never challenged a decision made by the local officials. Mac had opened his first welding shop in Evergreen in 1939 and until the spring of 1956 had never had any problem in the operation. The only problem that he was facing in 1956 was caused by other party's pride that was supported by egos and gave birth to vindictiveness that only became apparent and was exhibited when they were in a group. They were seeking vengeance for his having led the zoning effort a few years prior to his attempt to construct a new building and relocate.

As Mac expected the City Council refused to restore the building permit, so Jewel and Mac made an appointment with an attorney in Atmore, Alabama who, after listening to the story made a telephone call to the Unite States District Attorney in Mobile followed by a call to the Mayor, Pro-Tem in Evergreen. He then advised Jewel and Mac to return home and expect to hear more in a day or two.

The next morning Mac was called to a meeting in the Evergreen City Hall and soon returned with a new Building Permit. Jewel and Mac built two additions on the shop and seven houses in the little town after that confrontation and never had a repetition of the problem with any City Officials.

In June 1956 David returned to the Auburn campus, but instead of living in, "Mongolian Hole" he shared a room with another student in a converted garage at 143 ½ W. Glenn Ave. known as, "Mrs. Harmon's House of Handsome Heroes". He and his roommate were not the only residents at that address.

He did not renew his association with any members of the Auburn Sports Car Club and his academic performance improved considerably. He did learn that Bob Mullins and Fran Cunningham, both from Birmingham had entered into an arrangement with the farmer that owned the property formerly known as Sharpe's Field and were operating a drag strip on the site with motivation being personal profit. That assured the demise of the ASCC.

The course that engineering students dreaded and told horror stories about was American History. David had to take it that quarter and he was again reminded of Mr. Harry Dey, his High School History teacher. He earned a "B" in American History that quarter.

When the quarter ended he was told that the "co-op" employment for the next quarter would be with Combustion Engineering, Superheater, Inc. in Chattanooga, Tennessee. After a week at home in Evergreen Mac gave him twenty dollars, shook hands, and he rode the, "Greyhound" to Atlanta and then Chattanooga where he knew absolutely no one.

On arrival in Chattanooga on Sunday morning he rented a locker to put his trunks in, purchased a newspaper, marked a few advertisements for room and board, and started making telephone calls. That led to his hiring a taxi to go to a house on Duncan Avenue and renting a room where he would have a roommate, Lester Smith who was a heavy equipment operator.

On Monday morning he caught a bus to downtown Chattanooga, changed busses, rode to a bus stop that the driver told him was nearest his destination and walked the final three quarters of a mile to Combustion Engineering, Superheater, Inc. and reported for work. He was led through the darkest, dirtiest production facility he had ever seen. It was a boiler factory and he was assigned the position of, "Maintenance Electrician's Helper" that ultimately led to his work assignments causing him to eventually see the entire facility.

They had rules. An employee reported for work at 7:00 o'clock in the morning, worked without a break until 11:00, was allowed fifteen minutes for lunch, and then worked without a break until 3:00 o'clock when they went to the showers, changed into street clothes, and then left the site. One of the rules was, an employee was not allowed to sit down except during the fifteen-minute lunch break.

His assignment was to assist "Pap" Herndon in overhauling electric welding machines which he soon began to feel competent in doing, but he noticed that if you were working on one of the overhead cranes mounted on rails forty feet above the floor you were allowed to sit down. Somehow, Moody Gault, the Electrical Shop Foreman assigned him to maintenance on the cranes. He could sit down while working on anything located at a height. They recommended it.

Another assignment that he participated in was the assembly of some automatic wire fed welding equipment in the, "nuclear bay". The nuclear bay was the location for the construction of the nuclear powered pressure vessels for an electric power plant somewhere in Pennsylvania. He was assigned to be the assistant for an electrician known only as, "Paul". Paul was obviously an outstanding technician and electrician, but he did not change his coveralls, shower, or shave regularly. In the one and a half months that David worked with him he cleaned up and changed once. The other electricians in the shop wrote a note on the calendar with red crayon that stated, "PAUL CHANGED" and circled the date.

His personal life was slightly better than his working life. On Friday evenings he'd attend a movie and on the other evenings his landlady and landlord allowed him to accompany them in their living room and watch television, but he enjoyed reading in his room equally as much.

His landlady and landlord, the Barons were friends with another couple, the Moyers who bought a wrecked '56 Ford station wagon and borrowed use of the Barron's garage to use as a working space for Jake Moyer to repair the car and David started spending his evenings in the garage with Jake and the car resulting in a friendship that he enjoyed for several years.

His assignment for the quarter came to an end in the third or fourth week of December 1956 and he spent the Christmas holidays in Evergreen and then returned to Auburn and, "Mrs. Harmon's House of Handsome Heroes".

David registered for a sixteen credit hour load in college for the Winter Quarter of 1957. He was finally beginning to learn something about ambition sometimes promising more than reality can deliver.

Several of the students with whom he had become acquainted in the summer quarter of '55 while residing in Division "O" of "Mongolian Hole" had become members of the social fraternity, Tau Kappa Epsilon and after being invited to and attending a few parties at the Fraternity House he was asked to "pledge" TKE which he did. That is, he was asked to consider joining. He accepted and "pledging" means he started studying the history, rules, and so forth while those already members became more acquainted with the "pledge" and made sure his academic status met the membership requirements. His activity around the, Teke House was the only social activity he participated in that quarter and his academic scores were acceptable.

In March 1957 he returned to Chattanooga and Combustion Engineering where his job assignment was to be a, Foundry Maintenance Mechanic's Assistant.

During the preceding quarter he had maintained correspondence with Ruth Moyer, Jake's wife and had agreed to live in their home if he returned to Chattanooga. The Moyer's home was a single story ranch with an attached garage and the attic was converted into an air-conditioned bedroom that David shared with their ten-year old son, Francis. There were three bedrooms on the main floor, one of which was the master bedroom, the older daughter, Celeste had one, and their two younger daughters shared the other. It was the first time in his life that he resided in an air-conditioned home. Even the dorm at Auburn, "Mongolian Hole" was not air-conditioned.

At Combustion Engineering he was immediately assigned to the afternoon shift that started at 3:00 o'clock in the afternoon and ended at 11:00 o'clock in the evening. The assignment may have been made to, teach this "swingin' college kid" something and the rule against sitting down during working hours that he had dealt with in the electrical maintenance shop still applied.

During his first evening working on the second shift he met a mechanic who lived in Ringgold, Georgia and commuted through East Ridge, the suburb of Chattanooga where he then resided and made arrangements to be picked up beside Highway 41 in the afternoon and dropped off at the same point in the evening. "Cut" Sively was probably in his fifties and was very quiet. He was so quiet that he was even intimidating in his manner. It was suspected by some of the personnel in the Foundry Maintenance Department that "Cut" had earned his nickname during some of his social activities when he was younger. David was careful to avoid creating any reasons to confirm the rumor.

Combustion Engineering operated two foundries in 1957 and the "new foundry" was the one that David was assigned to. It was assumed that the maintenance staff would report to the site every Saturday morning at 7:00 o'clock and work until the maintenance and repair jobs were completed. Of course, they had already worked forty hours from Monday through Friday and the Saturday sessions would, by necessity run a minimum of eight hours long and it wasn't unusual for them to run twelve hours long. The compensation was paid at the rate of "time and a half" for those Saturday sessions and David was banking the money.

The working conditions were dirty and hot with dust from the casting sand creating haze that made it impossible to see from one side of the building to the other and because the molten iron was being transported from the cupola to the automatic casting machines in open ladles, poured into holding ladles on each machine's loading end, and the finished cast iron pipe or fittings produced were then hung on conveyors and moved around the building while cooling. The coolest ambient air temperature on the second floor of the new foundry during that quarter in David's career was one hundred twenty degrees.

One night per week his supervisor would give him a ride home and they'd stop at a drive-in restaurant on the way, drink one beer, and then go home. After an eight-hour shift in the dirty building at more than one hundred twenty degrees the shower and change of clothing in the bathhouse followed by one cold beer was G-O-O-O-D.

He soon met a young man working as a mechanic in the foundry named "Freddy" Friar who was able to get Friday nights and Saturdays off from the job. He was a stock car racing driver with quite a reputation and had an older brother, Harold who was even better known in the area as a driver. It wasn't long before David learned where the Friar brothers maintained their cars and started visiting their shop occasionally. He never understood why they'd make him feel welcome in their shop, but he still enjoyed the visits. He soon learned that Chattanooga was a real, hot bed of modified automobile activity, but his objective was a college education, so he maintained his self control and avoided getting involved in the auto racing activities.

He also noticed that a lot of teenagers were driving older Ford cars that sounded as though they had later model engines in them. Yes, there was an enterprise in Chattanooga called, "ridge running". It was transporting illegally produced alcoholic refreshment in cars that had been modified specifically for that purpose and unlike the popular legends, one of which became the, car culture movie, "Thunder Road" the drivers didn't attempt radical and dramatic escapes from the, "revenuers". They might on some occasions attempt an escape for a short time, but would soon turn into a side road or street and abandon the car with the "bootlegged booze" still in it.

The authorities would tow the car to the City Garage in downtown Chattanooga and store the car until they had enough captured cars to hold an auction and apparently some of the cars were sold relatively inexpensively. David never pursued that story and never met anyone who could confirm it.

David decided that he should keep a supply of beer in the Moyer's refrigerator for his rejuvenation and "re-hydration", but there was a requirement to financially support such an endeavor without compromising his primary objective. Jake understood all of the problems and volunteered to guide and assist the needy student in resolving the dilemma.

Jake had a bottle capper and an earthen crock of five gallons capacity. In a short time Jake had a supply of bottles, bottle caps, and ingredients specified by the recipe.

Ruth decided that it was time for her to use the Easter Holiday to visit her original home and family in the vicinity of Columbia, South Carolina, so she and the four children, Celeste, Lola, Francis, and Patsy departed in the Ford station wagon with plans to be away for at least a week.

Jake and David promptly transported all of the equipment and the recipe ingredients for their brewery into the kitchen, set up the operation, and within one evening had the earthen crock filled with fluid, safely located in the pantry, and all signs of any activity eliminated.

Three days later they bottled and capped the fluid from the crock into the bottles and placed the bottled product into cardboard boxes and stored those boxes in the rafters of the garage. Everything was under control and doing well until Easter Sunday afternoon when they were standing at the fence in the backyard separating the Moyer's yard from the neighbor's property and engaged in conversation with the neighbor.

They heard an explosion in the garage. In a short time they heard another explosion followed by another one. Jake and David excused themselves from the conversation, went into the garage and could see fluid dripping from the rafters and running down the walls.

Jake, originally from Pennsylvania's Dutch country said with his noticeable accent, "I ting ve bottle 'em too soon. Ve better get 'em down und unbottle 'em und ve better vouldn't do zat n-o-o-o-o more."

They used great care in lowering the boxes of bottled brew from the rafters to the floor of the garage and then moved them into the kitchen, set the empty crock on the table, and wrapped a bath towel around each bottle as they carefully lifted it from the box, untapped it, and attempted to let it empty into the crock. They had obviously bottled the brew before fermentation was completed.

When Jake removed the cap from the first bottle he found that the internal pressure was extremely high. Homebrew sprayed all over the kitchen as he attempted to direct it into the crock. He repeated that disastrous maneuver several times as he transferred the brew that had not exploded in the garage attic into the crock.

They washed the kitchen walls, cabinets, floor, and anything else that had the residue from the explosive brew on it that *they* could see. David had heard that, "Hell hath no fury like a woman scorned." He also remembered his mother, Jewel and her exhibitions of dissatisfaction with some developments when he was still at home. After observing Ruth's response to her discoveries in the kitchen, the pantry, and the adjoining dining room, and they went on for days after she returned from South Carolina he decided that he never wished to encounter a scorned woman.

One evening shortly after Ruth's return he arrived from the foundry and noticed that the kitchen light was on. He investigated and found that Jake was sitting at the table alone. Jake explained that he had found Ruth angry when he arrived home after working all day.

Sometime during the day Ruth had gone into the garage where she kept all of her potted plants during the winter to protect them from the weather. She placed them on the ties between the studs that made up the framing for the exterior walls of the building and when the bottles of brew in the attic of the garage exploded some of the fluid leaked down the walls and into the potted plants. Ruth was certain that the homebrew had killed her potted plants.

Inspection revealed that they exhibited the result of some type of trauma. Their leaves were hanging limply and appeared to be wrinkled and shrunken, but no part of the plants appeared to be dry or brittle as might be expected if they had died. Ruth wouldn't touch or look at the plants for days and then she decided to resume watering them. In two or three weeks they began to appear to be recovering from whatever had affected them.

Jake and David asserted that the plants were only suffering from hangovers. Ruth's disposition improved too.

Everyone acquainted with David was aware of his interest in automobiles and his fascination with hot rods, sports cars, and racing. He had retrieved a Model "T" frame from the same scrap pile that the components for the, "Jimmy-Wayne" flying machine had been retrieved from and had purchased a front axle, and a rear axle, from a '40 Ford from a friend in Evergreen and started to assemble "something" at Mac's Repair Shop.

Jake had a lot of "contacts" in the automobile business around Chattanooga and when a '57 Ford was burnt, declared a total loss by the insurance company, and was being scrapped Jake helped David acquire the 296 cubic inch displacement engine from it.

Ruth and Jake allowed him to disassemble, clean and inspect all of the parts, hone the cylinders, lap the valves, replace the piston rings, all of the gaskets, repaint the exterior of the major components, reassemble, and store the engine in their garage when he returned to Auburn for the summer quarter.

He had selected the TKE Fraternity House on E. Magnolia Avenue for his residence during the summer quarter of 1957 in Auburn. The house was an older private home that had six bedrooms on the second floor and at some time in its history had been a motel with at least a dozen rooms in one long structure in the backyard. The main floor of the house contained the original living room, a parlor with a hi-fidelity console in it on the right as one entered the front door, a dining room behind the parlor, a kitchen behind the dining room, a hallway from the living room extending rearward through the house to the back porch, and the House Mother's suite on the left of the hallway.

The fraternity managed it in a manner that assured cleanliness and safety, but it still had the appearance of an older home that was destined for destruction in the not far distant future and the cabins in the rear exhibited the condition of being very nearly unsuitable for occupancy. Condemnation of them would have been justified and appropriate. During the hot afternoons and evenings of that summer the fraternity members residing in the "cabins" had their electric fans running and placed lawn sprinklers on the metal roof in an effort to obtain some relief from the heat.

The summer quarters at Auburn were pretty quiet with perhaps twenty-five percent of the college's enrollment in attendance and the 1957 summer quarter was no different from any other summer quarter. In spite of the class schedule, studying, but not as much as he should have, and attending a movie every Saturday evening David didn't have anything else going on. The other students in the fraternity house existed in the same environment, of course. When the quarter was nearing its end the other students in the fraternity house placed a satellite speaker from the high fidelity sound system located in the parlor below David's room under his bed and routed the wiring from the speaker out the window, down the exterior of the house, in the parlor window, and connected it to the Hi-Fidelity set.

At approximately 1:00 o'clock the next morning David was aroused from his slumber and dreams by the very high volume of a recording by the, "Ted Heath Orchestra". There were no lights on in the house, but David knew that the "villains" had to be some of his friends in the house or from the cabins in the back. He found the speaker, disconnected the wiring, went back to bed, and resolved to never mention the event.

During his last evening in Auburn for that quarter he visited every bed in the TKE House and left some "itching powder" between the sheets. That evening he noticed that the showers seemed to be used more than usual and he found it difficult to sleep because of the low volume rumble throughout the house that bore a resemblance to complaints about some type of insects in the beds.

He never did or said anything to anyone about the "Ted Heath Orchestra" recording or the insect infestation in the beds.

He visited his family in Evergreen for a week and then rode a "Greyhound" back to Chattanooga where he began another quarter employed by Combustion Engineering. The position from early September through December 1957 was in an engineering office and he didn't know or understand what his duties were or should have been. The supervisor, Jack McClure seemed to resent having two "co-op" students imposed upon him and really didn't appear to have any wish to hear from or about them. The other student was from Georgia Tech.

David had moved in with the Moyers again and resumed the same social life that he had during the prior quarter in their home except one of his co-workers at Combustion Engineering and he had became good friends. Don Field's hobbies were riding motorcycles and chasing girls. He was proficient at both hobbies.

David enjoyed going places with Don on his Triumph bike, which was a 750 or 700 cc, two cylindered machine that was a serious motorcycle at the time. Don had a special transmission gear selector installed in order to ride because he had an artificial lower right leg and foot that eliminated the possibility of making gear selections with motion in the right ankle. He had lost his right foot and lower leg in a motorcycle-automobile collision when he was sixteen years old and the financial compensation that he received from the automobile driver's insurance company supported a life style that he enjoyed very much. Wayne Bridges was another co-worker in the office and resided in Ringgold, Georgia and was David's transportation provider between home and the job site.

An elderly gentleman from Indiana had retired and was towing a small house trailer with his '50 Plymouth tudor sedan to Florida to begin his retirement, ran a traffic light, and during the ensuing collision that demolished the front of the car, the house trailer broke loose from the trailer hitch and wrecked the rear. The accident occurred in front of the shop where Jake Moyer was employed.

Jake suggested that David purchase the wrecked Plymouth and repair it in the Moyer's attached garage during his spare time at home.

After agreeing on a price of forty-five dollars David bought the vehicle and moved it to the Moyer's garage where he spent all of his spare time for the next two months successfully repairing and repainting it. Christmas approached and David's "co-op" assignment came to an end. Ruth wanted to take the children to Columbia, South Carolina again and it was agreed that David would take them in his car, visit Ruth's family, and then drive to Evergreen to spend the Holidays with his family. Jake would retrieve Ruth and the children after Christmas.

David enjoyed the visit in South Carolina and on the way to south Alabama he visited an uncle in Augusta, Georgia whom he really didn't know very well at the time, but later came to know and consider him to be his favorite uncle, Edwin E. Currence, his mother's brother whose middle name he shared had made a thirty year career in the U.S. Army.

In Evergreen he had to admit to his parents that he hadn't saved money during the last quarter in Chattanooga and the Plymouth was his only acquisition during the past three and a half months. It couldn't contribute anything to his college education, but it was contributing to his education in life and reality. He had displayed a lack of wisdom, had shown immaturity, lack of discretion, impatience, poor financial management, no self-control, and no judgment. He couldn't claim that he'd been misled. He had known what he wanted and he had allowed that desire to take precedence over common sense and self-control. There would be a penalty to pay and he knew it.

Mac and he visited the bank, borrowed three hundred dollars with repayment required in twelve months, David was the borrower and Mac was the co-signer of the note. In January David returned to the Auburn Campus residing in the Teke House Annex on East Magnolia Avenue. While he'd been in Chattanooga the fraternity had purchased the house on the west side of the original Fraternity House and in spite of the "Annex" having been abandoned and allowed to deteriorate to the degree that it shouldn't have been occupied, the fraternity was using it as rooms for members.

He attended all of his classes, but his heart wasn't in it. He simply had no hope or confidence at that time or reason to believe that he'd ever hold an engineering degree and he quit trying. His only social activity was with the fraternity where he was "initiated" into the organization after tolerating the weeklong period of abuse called, "Hell Week". One of his old ASCC friends also became a "Teke" that week and lived in the "Annex", but not in the same room.

Bruce Lund owned a '55 Ford Thunderbird at the time and had replaced the 296 cubic inch engine with the 312 cubic inch, '57 model year engine and equipped it with a McCulloch supercharger. He took David with him one Sunday to the Montgomery Speedway and Drag Strip, but neither of them seemed to enjoy it. The car didn't perform well in practice runs, so' Bruce chose not to compete in it.

The greatest distraction to his studies though was the result of his agreeing to do a favor for one of the fraternity members who had a Chevrolet sedan that suffered an engine failure. David attempted to overhaul the six-cylinder engine in the car parked in the backyard of the Teke House Annex using tools borrowed from Bruce Lund and lapping the valves into their seats in the cylinder head on the desk in his room. The overhaul proved to be a futile effort.

The winter quarter, 1958 ended with David feeling frustrated, defeated, depressed, hopeless, and despondent, so he went to Evergreen for a few days and then returned to Chattanooga to seek employment. He was afraid to advise Jewel and Mac of his lack of employment.

The financial recession of 1958 was at its peak or lowest level depending on how one perceived it and the primary financial source for David was from small things like in the northern part of the city of Chattanooga working for a business called, The Lakeway Garage. The two guys that owned it had totally filthy personal hygiene habits and a vocabulary as dirty and rough as the garage floor. David had worked in some jobs that were normally held by some crude and unrefined people, but he had never been associated with people as bad as those two guys. He stayed there for two weeks and quit.

He then loaded the Ford engine that was stored in the Moyer's garage into the trunk of the '50 Plymouth sedan and his personal possessions into the rear seat and returned to Evergreen the second week in May 1958 where Mac put him to work in his welding shop.

'50 Plymouth with '49 front end (author's collection)

Hot Rod project with '56 Ford engine, '40 Ford running gear on Model T frame (author's collection)

Three years had passed since his graduation from High School and departure for college. In that time almost all of his acquaintances had found employment in other locations, married, gone to college, or joined the armed forces. His opportunity for a social life was almost nonexistent and one could not have been supported on his earning capability anyway. He tried to reestablish the contacts that were made while he was in High School and seeking employment, but all of those doors were closed.

During this time period he tried to make some progress on the hot-rod/racing car he had been trying to build, but without money to support the project he couldn't make any progress, so he started modifying what he had. Those changes weren't being made with an objective in mind. They were being performed to, "burn time" and eventually led to the abandonment of the project. He sold most of the parts that were being used and certainly didn't lose anything financially.

Jan Hendrix decided that he needed an automobile and that led to David driving him to Montgomery for a shopping trip. Jan purchased a '57 Renault 4CV sedan, but he was not comfortable while driving it and the two of them decided that David would drive the Renault back to Evergreen and Jan would drive the Plymouth. The Renault was nearly new and had a fast ratio steering gear and being lighter it also had lower steering efforts. The '50 Plymouth was soft riding, slow steering, and, "cushy" by comparison even though it accelerated faster and had a significantly higher top speed.

Later that week David was directed by Mac to visit the Engineering Department at Southern Coach and Body Company, Inc. and ask for the Chief Engineer, Edsel Williams. He was invited into a small room that was the home of the Engineering Department and asked to demonstrate his drafting ability by performing the detailing of a small job. The visit resulted in an employment offer at the rate of one dollar and thirty-five cents per hour.

The interview had been the result of the friendship between Mac and W. C. Bowers, the Vice President of Engineering and Manufacturing at Southern Coach and Body Co.

PART XV

SOUTHERN COACH & BEGINNING A CAREER

Southern Coach had its origin in Mr. Watson's sheet metal shop on Carey Street, later renamed Martin L. King Street in Evergreen during the 1920's with the construction of a school bus body on a Model "T" Ford truck chassis and was called the, "Watson-Brown Company" that was sold to Stanley Green, Allard C. French, Sr., and Ceylon P. Strong to become Southern Coach Manufacturing Co. in June, 1941.

Allard French and Ceylon Strong designed the integral type transit bus that was produced for use in cities throughout the United States and sometimes modified for use in some cities located in South America. Production of the transit bus was the Company's, "bread and butter" until it became necessary to produce the forward control, walk-in style delivery truck body in 1956 to maintain survival.

In 1958 seven gentlemen combined their resources, purchased the company, and Southern Coach & Body, Co., Inc. was organized with Clayton Albert as President, W. C. Bowers, Vice President in charge of Manufacturing and Engineering, Charles E. Scott, Vice President in charge of Sales, Mack Williams, Production Superintendent, and Delma Bowers, Welding Foreman, with Elmer Padgett completing the management team. Herman Bolden remained on the site and was instrumental in making the transition from Southern Coach Manufacturing to the new organization.

The first products were carry over designs of forward control, walk-in, delivery truck bodies built on G.M.C., Chevrolet, Dodge, and Ford truck chassis. Those bodies were sold in three basic designs called, All Steel, Steel Frame with Aluminum Skin, and All Aluminum with cargo compartments varying in two-foot increments from ten feet long to sixteen feet long. It was a continuation of production begun when the organization was still called, "Southern Coach Manufacturing Co.". It seemed like every sale required the Engineering Department to provide production drawings for a body that had been advertised and sold, but had never been designed. Edsel Williams and David McKenzie were busy people producing two designs per week, running copies of the drawings, and going to the shop to look at problems and answer questions.

The job wasn't driving racing cars or designing automobiles, but designing truck bodies was as close to those objectives as David believed he was likely to get. He was resigned to his fate, his profession, and was ready to visit his friends in Chattanooga after being employed at Southern Coach and Body Company, Inc. for one week.

On Thursday afternoon, July 3 Jan Hendrix and David McKenzie left Evergreen in the Renault 4CV sedan with Jan planning to drive until darkness overtook them. Then they'd change drivers and David would complete the trip to Chattanooga. David slid down in his seat, placed his knees against the instrument panel, pulled the black and white checkered English golfing cap over his eyes, and began to doze.

There was an unusual movement of the car that awakened David and with one hand he lifted the cap and saw that the vehicle was, fishtailing its way around a curve to the right while going downhill and the road was wet and slippery. He said, "Stay off the brakes." Jan stood on the brake pedal, quit wrestling with the steering wheel, and the car crossed the road and began to overturn as it started down a creek bank sideward. It completely rolled over to the left for four hundred sixty degrees, struck the opposite bank of the creek simultaneously striking a dead tree stump that was inhabited by a nest of hornets, then rolled one hundred eighty degrees to its right, struck the bottom of the creek on its right side and skidded to a stop. When it struck the opposite bank of the creek and began rolling to the right David realized that he could see the rear of the right headlamp fairing and knew that his head was outside the window. He pushed as hard as he could with both hands and as the car was sliding along in the bottom of the creek he could hear an odd scraping noise.

When the vehicle stopped David was on his hands and knees on the right front door interior. His left hand was on the headliner above the door and the right hand was n the door post between the front and rear door, (It's called the "B" pillar in the auto industry.) and Jan was lying on his (David's) back. He asked, "Jan, are you O.K.?"

There was a mumble from Jan.

He asked again, "Are you hurt?"

Another mumble was the response that led David to believe that Jan wasn't seriously injured. Jan seldom spoke more clearly or louder.

David then said, "Well, get off my back so we can get out."

They climbed out the left front door of the car, jumped from the car to the nearest bank of the creek and were met by the swarm of irate hornets that encouraged them to wade across the shallow creek, climb the bank, and join the gathering crowd of passersby on the shoulder of Highway 31, eight miles north of Evergreen.

Jan Hendrix's Renault in the creek eight miles north of Evergreen on U.S. Highway 31, July 31, 1957 (author's collection)

Jan was driven to a hospital in Greenville that was operated by his mother's family, the Spears. Someone contacted the Highway Patrol and David was given a ride back to Evergreen by the Patrolman who had arrested him two weeks previously for speeding. Neither Jan or David had been seriously injured in the accident. David, seven hornet stings, and the top of his head scraped open on the bottom of the creek, suffered the most serious physical trauma. Apparently the hornets really objected to the intrusion of the little French automobile into their home, the stump. That would have been consistent with Mark Twain's opinions of the French

Jewel expressed some doubt about the wisdom of attempting to operate a small European vehicle designed for that environment on the roads in the United States. She didn't say very much about it to David and Mac didn't mention it, but David heard some comments about Jewel's observations and recommendations from a few family friends.

At this time in his life his parents used David when addressing him. When in High School the other students started using his middle name. He did not approve of the presumptuous students assuming the liberty and right to rename him, but he never expressed an objection. Upon returning to Evergreen after the three-year separation from the community he had started using the name, "Dave".

The order for the S-45-DM busses had been completed and the Engineering Department staffed by Edsel Williams, Roy Kendrick, and Dave McKenzie was producing three truck body designs every week while the Production Department was completing seven and sometimes twelve bodies per week when in January or February of 1959 Charley Scott suggested that the company produce a small delivery truck utilizing unitized construction. It would not be the traditional body on chassis configuration that the company had been producing and since the size was not limited by a "purchased chassis" as had been the case up until then, the opportunity was available to design and produce a NEW vehicle.

Edsel Williams, the Chief Engineer did the initial concept drawing and Roy Kendrick designed many of the major components while the crew in the shop disassembled a new Chevrolet pick-up truck to obtain chassis components. W. C. Bowers, Gene Majors, Delma Bowers, and Cleve Garner were primarily responsible for assembling the little truck and "Pee Wee's" father "Red" Middleton produced the sheet metal parts. The crew supervised by Dempsey Coburn primed and painted it. "Pee Wee" was also employed by Southern Coach and his automobile was a new Oldsmobile hardtop. He fit the image of the recent High School graduate starting life in south Alabama in 1955.

Concept, design, and construction of the prototype vehicle had been accomplished in TWO WEEKS and it was then shown to the public at a truck show in Minneapolis, Minnesota where executives from General Motors Corporation saw it and negotiations soon began in Michigan for Southern Coach to manufacture the vehicle which would use G.M.C. instead of Chevrolet chassis components and would be marketed through GM dealers.

Southern Coach soon had a contract with General Motors Corporation to produce the Junior Van. The name, "Junior Van" was chosen to celebrate the birth of a G.M. executive's son. Southern Coach did not have the production capacity to meet the G.M.C. contract requirements.

On the east side of Rural Street was an abandoned collection of buildings, which had been used as the site of a plant that produced Venetian blinds during World War II. Of course, supporting the establishment of an automotive production line in that facility presented a financial challenge that was resolved by an investor named Elrod loaning Southern Coach enough money to refurbish the buildings into a usable condition. The loan was secured by the City of Evergreen passing a Bond Issue to guarantee repayment and restoration and modification of the, "Venetian Blind Factory" was begun that led to production of the Junior Van at the plant renamed, The Elrod Plant in 1960

It is remarkable how the financial tactic that was used in south Alabama during 1959 resembled and shared similarities with the methods used by Lee Iacocca in 1982 to save Chrysler Corporation from bankruptcy and failure. The Detroit based auto industry leaders discovered Dr. John Demming in the late 1980's or early 90's and espoused several of the management concepts that he was teaching after studying competitive operations in Japan. One of those was called, "Concept to Customer in two years". "Concept to Customer" hadn't required two years at Southern Coach.

No one knows or has heard of the creative and innovative tactics originated by a small group of unknown and unrecognized self starters in an agrarian part of the country thought of generally as nonproductive and occupied by a lazy populace. Ignorance is bliss and performs an excellent job of supporting egos.

The employment in the Engineering Department at Southern Coach was very satisfying and even though his salary was low there was considerable psychic reward in his duties. It rejuvenated his confidence and hope that there was potential for success in life, but he had no social life and no contemporaries in Evergreen. There was nothing for him to do and no one for him to associate with and then he heard a rumor about some possible interest in flying and an organization named the, "Civil Air Patrol."

PART XVI

ON THE WAY UP AT LAST

Ivey T. Booker and Lee F. Smith had been attempting to gain the attention and support of the Alabama Wing, Civil Air Patrol and hadn't been able to obtain a serious response until John Patten, then a High School Senior had, while on a school sanctioned tour of the State's Capitol, Montgomery visited the State Department of Aeronautics and discussed the issue with the Department Director, Asa Rountree who interceded on behalf of the aviation enthusiasts in Evergreen, and encouraged the response from the Civil Air Patrol, Alabama Wing Headquarters in Birmingham.

A meeting was held on the second floor of the Conecuh County Courthouse that Dave attended and was surprised when he saw his father in the room. He suspected that if Jewel hadn't already tied his father in the house she'd kill him when he returned. During the meeting it soon became apparent that the group of twenty five or thirty people attending were hoping to form a CAP Squadron and be rewarded by the United States Air Force providing them an airplane and teaching them to fly while also maintaining and supporting the airplane.

As the meeting progressed it became obvious that the CAP wasn't going to teach anyone to fly and certainly wasn't going to give a group of Student Pilots an airplane. There was noticeable disappointment exhibited by attendees at the meeting. Dave thought of his mother's appraisal of aviation enthusiasts.

Mac McKenzie then suggested that the group organize a flying club and the mood in the room changed again resulting in ultimately forming the Evergreen Composite Squadron, Civil Air Patrol and the Conecuh County Aero Club. Mac didn't join either organization, at least not then.

Ivey Booker and Bobby McLendon were taking flying lessons at a small airport located on the south side of Montgomery named, "Allenport" which was owned by Dr. Wren Allen and it was on that field that the men who formed the Conecuh County Aero Club found two airplanes.

There were twenty-six enthusiasts who invested twenty-five dollars each into forming the Aero Club and five of them formed their own partnership. The two groups then bought two airplanes at Allenport. One was a 1939 Aeronca Defender, Model TAC supposedly blessed with a good airframe, but handicapped with a tired engine. The other machine, an Aeronca Chief, Model 11AC allegedly presented the opposite situation, good engine, poor airframe. Five of the investors purchased the one with the good engine and all of them purchased the one with the unserviceable engine. A separate agreement was made among the twenty-six investors that led to an engine swap that yielded two airplanes. One had both an unserviceable airframe and engine and the other with supposedly airworthy components. In other words, "One airplane requiring restoration and one good one."

After all the negotiations and the engine swap were completed, the five guys with the rebuildable airplane had a total investment of one hundred fifty dollars in their project and the twenty-six members of the flying club had a total investment of six hundred fifty dollars in the flyable airplane.

Several of them devoted an entire Sunday, September 19, 1958 at Allenport to performing the engine swap, disassembling the airplane needing restoration, and loading it onto a truck trailer that had been borrowed from R. V. McLendon and that evening Bobby McLendon , John Patten, and Dave McKenzie returned to Evergreen in the truck with the Aeronca Chief on the trailer.

Somehow, Dave learned that a flight instructor from Pensacola was going to visit Middleton Field and was seeking students, so on September 24, 1958 he took an extended lunch break, met Ross Shuel, and took his first forty-five minutes of dual instruction in a Cessna 140. After leaving the office at the end of the day he returned to the airport and was fortunate enough to find that Mr. Shuel hadn't departed and took another forty-five minutes of flight instruction. He loved it.

It was true at that time and is still true that sixteen years old is the required age to legally solo an airplane and seventeen is the age required to earn a Private Pilot Certificate, but those achievements require the parental approval if the pilot is less than twenty one years old.

Dave knew that there was no way to acquire parental approval from his mother, so he had been forced to wait until he was twenty one years old and legally designated an adult before starting flying lessons and on September 27, 1958 he took another lesson at Auburn, Alabama from E. E. Jones in an Aeronca Champion Model 7EC that lasted twenty five minutes. He was committed. He then had one hour and fifty-five minutes total student pilot time and experience logged. He was on his way to his future.

He certainly didn't plan to tell Jewel what was happening and he thought that if Mac learned of it and didn't tell Jewel there would probably be two seriously injured males at 115 Magnolia Avenue in Evergreen. He didn't know how he was going to handle the confrontation that was in his destiny, but it was his choice, his love, his passion, his life, and his responsibility. He was going to fly.

In the movie, "Tora, Tora, Tora", there is an act that occurs during the arrival of the Japanese bombers over the island of Oahu, Hawaii that shows the Japanese force encountering a light airplane flown by a woman and a young male student pilot who simply dove out of the way of the attacking planes. The truth is a father and his son were enjoying a Sunday morning pleasure flight in their Aeronca Defender and the Japanese shot them down. The occupants of the Defender survived the attack and it may be claimed that the first American casualty of World War II was an unarmed civilian light plane.

The Defender was a two passenger, high wing monoplane powered by a sixty five horsepower Continental engine which carried the passenger in the front seat and the pilot in the rear seat while cruising about seventy two miles per hour.

During the period of time that the acquisition of the airplane was taking place another gentleman from Brewton, Alabama contacted the group and offered his services as a flight instructor. W. H. "Bill" Blacksher had resigned from his position as a DC-4 Captain with National Airlines after landing at Jacksonville, Florida with a loss of hydraulic pressure that led to the airplane running off the end of the runway and causing him to unlock the landing gear to stop before running across a busy highway. He was trying to continue flying while seeking employment as a pilot. He and Mac had been acquainted at the Brewton, Alabama airport in '43 or '44.

On September 28 a flight instructor from Montgomery delivered the Defender to the Conecuh County Aero Club at Middleton Field, Evergreen, Alabama and on September 29 Dave had his fourth flying lesson that was his first with Bill Blacksher in the Aeronca Defender.

David knew that he needed a document called, Third Class Medical Certificate to fly an airplane solo and that he would not be allowed to solo until he held a "Medical". At the time, the CAA allowed any physician to perform a Third Class medical examination, issue the certificate to the applicant, submit the report of the examination to the CAA Aeromedical Branch, and the CAA handled it from that point, so in his travels Dave obtained a copy of the medical exam report form and approached Dr. R. L. Yeargen requesting an examination.

Dr. Yeargen asked for his glasses, peered through them, and said, "Give it up. You can't see well enough to fly and I won't take your money and I don't recommend that you waste it on this effort. You have no future as a pilot."

Dave took the form, departed, and made an appointment with Dr. "Bill" Turk who conducted the examination, asked for eight dollars, gave the youngster his copy of the document, and Dave enjoyed the sensation of relief. He thought he would fly.

On October 26 Dave was completing his eleventh lesson, the eighth with Bill Blacksher, and as they taxied back to the parking ramp he was startled to see Mac's Chevrolet sedan parked on the ramp. The fact that it was impossible to do anything in the little town of Evergreen without everyone knowing about it was reaffirmed and he began to think that Jewel probably learned of his activities and had sent Mac to force him to quit flying and return home. He suspected that there had already been, "war in the camp" and he would be the next casualty.

Bill Blacksher completed the logbook entry while conducting a conversation with Mac. Dave didn't know that the two of them were acquainted from the period when Mac flew the Bird biplane from Brewton. Charles Lindberg had said, "If two pilots don't know each other they'll both know some other pilot."

Dave had remained beside the airplane while Bill and Mac had stood beside an automobile that Bill was using the hood from for a desk and he was feeling great uneasiness as the two men turned, looked at, and then approached him. "Your dad wants to ride with you." Bill said as he handed Dave's log book to him with the latest entry adding up to a total time of seven hours and forty-five minutes of dual instruction logged.

Dave entered the rear seat and Mac got in the front seat. Bill "propped" them. That's to say that he turned the propeller by hand to start the engine. Dave taxied out to the runway and on the way he asked, "What do want me to do?"

Mac's answer was, "Stay in the pattern and show me some landings." Nothing else was said.

Dave went through the pre-takeoff checks, tookoff, flew around the traffic pattern, and landed. Mac waved one hand above a shoulder, which was a signal to takeoff again. Mac never said a word as they made four trips around the pattern and during the roll out phase of the last landing motioned towards the ramp.

Dave taxied to the ramp, parked, and shut down. Mac got out and started walking away as Bill queried, "What do you think?"

Mac replied, "He needs some more dual."

Dave thought, "There's going to be HELL to pay."

On Saturday afternoon, November 8, 1958 Bill Blacksher gave Dave one hour and ten minutes of dual instruction in takeoffs and landings, got out of the airplane, endorsed the back of his Airman Medical Certificate, wrote an entry into his log book, and told him to make three takeoffs and landings SOLO.

Dave had completed nine hours and twenty minutes of instruction given by three different instructors in three makes and models of airplanes and included a thirty minute evaluation by his father. He was, "one proud young man with no one available to listen to the story."

Aeronca "Defender" in the hangar on the site now occupied by the "Big Red" hangar (author's collection)

Preparing to fly after work in April, 1959. Bobby Stewart is assisting. (author's collection)

Pre-flight inspection being performed by Dave McKenzie on the "Defender" (author's collection)

Soloing is supposed to be a confidence building maneuver. It is supposed to demonstrate to the Student Pilot that he or she is able to fly the airplane unassisted. Admittedly, it is never a surprise to the student that he or she can accomplish this, but if the instructor does not allow the student to fly alone, the student begins to have self-doubts. This shows up in each lesson after the proper time to solo comes and goes. The instructor can see a regression or decline in the slope of the learning curve. Then it behooves the instructor to recognize the, "right time". If the instructor releases the student to their assignment too early disaster is almost inevitable. If the student is restrained he or she may never become a pilot. Think of the psychology used here and remember that for the sake of survival of both student and instructor, the psychologist-instructor cannot observe, or listen to the patient-student while attempting analysis. The only symptom available for use in analysis is the response of the airplane. The instructor MUST be correct. Bill Blacksher was charging three dollars per hour for his services. He was surely underpaid, but he was still correct in allowing

Dave to solo when he did. Dave's confidence was not improved by his performance, but it was safe and encouraging. He bounced on all three landings that he made.

On Sunday morning, November 9 Bill and another student were barely able to climb above the trees at the edge of a cow pasture neighboring the south side of the airport. They did manage to cross the trees, but lost altitude over the pasture and had to fly in circles in it until they gained enough altitude to cross the trees again and "hop" back onto the airport. They diagnosed their troubles as being engine related.

Repairs, namely a top overhaul of the engine were made by Bobby McLendon. It ran fine after he completed the job even though he didn't have a license as an Aircraft and Engine Mechanic. He simply knew how to do the work that was required at the time. He should have had the CAA (Civil Aeronautics Administration) license or been working under the supervision of a CAA Licensed Mechanic. Isolated as they were though, the fact that "it worked" was enough evidence to gain the satisfaction and approval of the group in the flying club.

After that series of events most of the other club members' desire to fly diminished and the availability of the Defender improved for Dave.

Bill Blacksher obtained a position as a Flight Instructor in one of the Civilian Contract Schools teaching U.S. Air Force Cadets to fly in the Lockheed T-33 jet trainer precluding his further availability, so Dave's next flight was on November 30 when Mac rode with him again through six landings. He was on his way, but what a way to start!

Occasional glances out the office window as Saturday, December 13 progressed revealed and confirmed the low hanging overcast. Dave already knew that the temperature was about thirty-five degrees, Fahrenheit and he had no idea what the spread was. (The difference between the current temperature and the dew point) and he didn't know that he should be alert for possible icing under such conditions. He was working overtime and was looking forward to going flying on his third solo flight if the weather permitted.

He was overhauling his automobile's engine at Mac's shop, so he had made arrangements with Roy Kendrick, a coworker in the office to drive him to the airport. Eventually all of the variables came together and suffered conquest and solution. He stood at the airport under a low overcast with it sometimes drizzling rain. Perhaps it was the heat of his enthusiasm that kept him warm in the thirty-five degree temperature as he pulled the airplane out of the hangar, performed the pre-flight inspection, and started the engine by hand propping. It had no starter or electrical system. He did not allow participation by Roy because Roy had no training in aviation activities and it was probably beneficial for Dave personally to practice the procedures anyway.

Due to the weather conditions he had chosen to remain in the airport's traffic pattern and perform successive takeoffs and landings. Aviator language would be, "touch and goes". The first trip around the pattern was uneventful and the second began smoothly enough, but while turning to the crosswind leg of the traffic pattern the engine quit running. He continued turning while reaching for the carburetor heat control, which, in the Defender was located on the left windowsill.

He was reminded of a Country and Western song, "If You Feel Like You're in Love, Don't' Just Sit There". When things started happening he didn't just sit there. Use of carburetor heat did not improve the performance of the engine. He did not run a magneto check because in that airplane, he would have to change hands on the control stick to accomplish that. He did cycle the throttle violently, attempting to get the engine to run as he continued the left turn, gliding then, and also on the downwind leg of the pattern. By that time, he was already thinking of making a crosswind landing on runway 10. He didn't quit pushing and pulling the throttle violently though and occasionally the engine would backfire and then run for a very brief period of time. It was probably building carburetor ice faster than the heating system could counter it and the backfire caused by the violent throttle usage cleared the ice from the engine's induction system.

There is a good possibility that he could have kept the engine running intermittently to complete the trip around the pattern, but runway 10 looked very inviting in spite of being crosswind. He decided to land on it. Having made that decision he turned to final for runway 10, closed the throttle, and the engine quit running. He continued to glide towards a landing. He landed crosswind, dead stick, and without incident.

He decided that he had achieved his objective and had "flown enough" that day. He waited for a few minutes and then restarted the engine, taxied back to the shed used for a hangar, put everything away and rode with Roy back to town.

Review of his logbook shows that he had completed a total of twelve hours of flying time, one hour twenty minutes of it and eleven landings solo. One landing was a forced landing, a genuine emergency.

In retrospect and with many years and hours of experience and twenty-twenty hindsight he must wonder what might have happened if he had enjoyed proper supervision. No doubt, he would not have been flying in those weather conditions at that experience level and definitely not in that airplane. Certainly, the whole series of events would not have occurred.

Perhaps it's an illustration that justifies the necessity to, "bend the system" sometimes to meet the circumstances and achieve the objective.

During that period Dave worked on other projects in the Engineering Department and at 3:35 on most afternoons he'd go to the airport where he'd either fly the "Defender" or simply loaf in the FAA Communications Station called, "Evergreen Radio" by pilots who wished to avail themselves of some verbal assistance such as checking weather reports while traversing the area.

Dave had no instructor or supervision and on his personal initiative obtained copies of the Federal Air Regulations and would read them every evening until boredom over came his ambition and he'd succumb to the temptations of somnambulistic oblivion.

One evening he read that he should have been carrying a Student Pilot Certificate with the appropriate endorsement on it from his flight instructor prior to flying an airplane solo. He also learned that Bill Blacksher was not aware of that requirement and was not an authorized flight instructor. In the waning years of the CAA they began to require biennial renewal of Flight Instructor Certificates and Mr. Blacksher was not aware of that requirement and had never received a renewal of his Flight Instructor Certificate. His flight instructing activities had been totally illegal, but he hadn't been aware of it.

Dave also learned that he should have passed a written examination on Civil Air Regulations consisting of ten questions selected by a CAA Inspector or a Designated Examiner prior to soloing an airplane. Of course, he had never done that either, so on Saturday, March 28, 1959 he drove to Milton, Florida, introduced himself to the Chief Instructor at the Milton School of Aviation, Chuck Prince, presented his logbook, and described his requirements.

Chuck Prince introduced him to a gentleman who had earned his Private Pilot's certificate the previous week and David boarded an Aeronca 7-AC which they flew to Bell Air Service at the Pensacola Airport and sought out the examiner, Odie Bell who inspected Dave's logbook, listened to his story, and said, "Complete this examination while I call Jacksonville." Jacksonville, Florida was the location of the District Office of the newly established FAA. The Federal Aviation Administration with retired General Elwood P. Quesada as the first Administrator had come into existence on January 1, 1959 replacing the CAA.

A few minutes later Mr. Bell returned, graded the ten true/false questions on the examination that Dave had completed and said, "Give me five bucks. Here's your Student Pilot's Permit. You must be doing O.K.. You haven't killed yourself, yet."

Dave accepted the document and didn't say anything more than, "Thank you." as he and the new Private Pilot returned to the Aeronca, started it by hand propping, waited for the flashing green light from the control tower, and by complying with the clearances issued by the light signals departed from Pensacola and returned to Milton "T" Field.

At the time there was no requirement to have a radio in an airplane to operate from any airport in the United States except one in New York and one in Chicago. At busy, controlled airports the traffic for NORAD (non-radio equipped aircraft) was directed by light signals.

Later that afternoon Chuck Prince rode around the traffic pattern at Milton "T" Field with Dave flying the Aeronca and after three landings said, "I have it."

Dave removed his hands and feet from the controls as Chuck turned to the right and then made a steeply banked tear drop shaped turn to the left which completed the turn-around, landed, rolled out to the hangar at the south end of the runway, signed the Student Pilot Permit, handed it to Dave, got out of the airplane, and said, "Do three take-offs and landings, then return. Don't ever let me see you do anything like I just did."

That flight was Dave's first legal solo flight and he already had logged ten hours and ten minutes of dual instruction and thirty four hours of solo time, but for the first time in his career as a pilot he was, "LEGAL" and resolved to be a fair haired boy, to live clean, and to have pure thoughts forever more.

He didn't mention any of the day's activities to Jewel or Mac.

On Mother's Day, Sunday, May 10, 1959 Jewel and Jimmy had been left in Cuba, Alabama visiting her mother and relatives while Mac and Dave continued south on U.S. Highway 11 to Laurel where they visited Mac's mother, Dave's grandmother. As usual on those visits Mac and Dave went out to the airport to see "Hesler" and Dave enjoyed a very pleasant surprise.

Mac was adamant in his thoughts about aviation safety and two of his personal requirements that Dave adopted and became almost as strongly opinioned on were ONE, spin training requirements should NEVER have been deleted from the pilot's training and experience syllabus and TWO, forced landings should be included in the syllabus as well.

Dewey, as Alton Hesler called him said, "Hess, I have an errand to run. Would you take David up and teach spins to him?"

Hesler promptly agreed, turned to David and asked, "Have you ever flown a Luscombe?" Dave explained that he had some "dual" from his father in one when he was twelve to thirteen years old and Hesler told him to "pre-flight that one". It was a 1946 Luscombe 8-A powered by a 65 horsepower Continental engine with the all metal wings on it.

They strapped in and someone else propped the little airplane. Dave had no difficulty taxiing out to the runway, performing the pre takeoff checks, and taking off.

Hesler said, "Take us to thirty five hundred north of the airport." Dave did as directed and after Hesler had him demonstrate some stalls Hesler commanded, "I have it." He then demonstrated a three turn spin, gave control back to Dave and directed him to, "Climb back to thirty five hundred."

They reached the assigned altitude and Hesler said, "Now, you do the clearing turns and if there's no traffic, you do one." That was followed by another climb, another spin in the opposite direction, and then a repeat of the same series of events again.

They had recovered from the second spin performed by Dave when Hesler said, "Let me show you what your daddy used to do with my little Aeronca." He verbally described what he was doing as he made the clearing turns, reduced the throttle setting and air-speed, followed by a sudden and extreme pull rearward on his control stick and a nearly simultaneous application of full rudder travel with one foot. The airplane raised its nose and immediately rolled. Prior to the airplane acquiring an upright wings level attitude Hesler quickly moved the stick forward and applied opposite rudder. The airplane had rolled three hundred sixty degrees and returned to a slightly nose high, wings level attitude. He asked, "What do you think? Do you want to do it?"

Dave's response must have been the one Hesler was hoping for because he said, "See? It's nothing but a spin going in the horizontal direction instead of straight down." They spent the next half hour climbing to thirty five hundred feet of altitude, performing clearing turns, then a three turn spin in one direction and a recovery, followed by a single turn snap roll in the same direction. Then climbing back up, repeating the clearing turns, stalling, falling into a three turn spin in the opposite direction and another snap roll in the same direction as the spin. Dave had never had so much fun. Of course he was still very bashful and extremely shy.

After landing Hesler's endorsement with his license number, 13099 in his logbook indicating that he had given Dave thirty minutes of dual instruction in spins became one of Dave's most treasured memories. That endorsement brought his total flying time logged up to a total of sixty two hours and twenty five minutes. Thirteen hours and five minutes of it was dual instruction.

No one, not even Dave had any idea what a magnificent world had been opened to him.

PART XVII

THE MILITARY OBLIGATION

In January 1959 Dave had received the well known letter from the local Draft Board and had made the trip via Greyhound to Montgomery for the two days of pre-induction written and physical examinations that he soon learned he had passed with a very good score on the written exams. At this point Gomer Pile would have said, "Surprise, surprise, surprise!"

At that time every male homosapien born in the United States was obligated to serve six years in the military service of his country. One means of meeting that obligation was to wait and be drafted into the Army for two years, then serve two years active Reserve followed by two years in the inactive Reserves. The second was to volunteer for three years and then serve three years inactive Reserves. The third was to serve six months active duty followed by five and one half years in the active Reserves. Dave selected the third choice and joined the Alabama National Guard in January 1959.

He knew that his six month long assignment to active duty with the U.S. Army was to begin on June 15, 1959 and some adjustment to his life style was imminent, therefore he decided to sell the '50 Plymouth and become a pedestrian for the next six months. It was not destined to work out like that. He attempted to sell the Plymouth and the only offer he could obtain was an offer to accept it as a trade in on a used '57 Plymouth "fordor" sedan. He didn't feel that he had any choice, so he visited the local bank and outlined his predicament, which was, his monthly income from June 15 through December 15 was going to be sixty-two dollars per month.

The officer at the bank wasn't afraid of the situation and approved the loan to be repaid at the rate of fifty dollars per month for the first six months and then increase to seventy-five dollars per month for the duration of the contract for eighteen more months. On the first of June Dave had the '57 Plymouth with nineteen thousand miles on the odometer that had truly been owned by, "a little old lady". Honestly though, one would not have pictured Mrs. Stowers as "the little old lady". She was attractive, perhaps middle aged, and the wife of one of the wealthiest landowners in Conecuh County. He was fortunate to have her car.

On June fifteen he stepped off the Greyhound at Fort Jackson, South Carolina to begin ten weeks of basic training as a, "grunt". At the old age of twenty two he found that he was one of the, "old men" in Delta Company, 5th Battalion, 5th Regiment, of the 7th Army and because of his experience in the ROTC program in college he was made a, "Squad Leader" in the 3rd Platoon. That appointment left him with no friends in the Army.

The Unit was made up of National Guard troops except for the cadre. The Company Commanding Officer, the Executive Officer, the Field First Sergeant, and three of the four Platoon Sergeants were all ex Airborne-Rangers and they had no appreciation for National Guard, Reserve, or what they referred to as, "short timers". That is when they didn't call them, "candy assed" or "mama's babies".

The cadre made it a point to train the troops hard for the ten weeks of "Basic" and the pressure was applied as the troop's boots hit the ground in the Company Area. The physical pressure and mental harassment caused three of the two hundred personnel to break down and be removed from the Company. The rest of them finished the program.

After five weeks of training the troops were told they were going to receive a, "week-end pass" thanks to the kindness, tenderness, consideration, and generosity of the Company Commander and the rest of the cadre. The "week-end pass" would begin at 13:00 hours on Saturday afternoon and end at 24:00 hours Saturday evening. They were ordered not to leave the downtown area of Columbia, South

Carolina, not to mingle with the residents, to stay out of pawnshops, not to gamble, not to chase any girls, and not to visit more than two bars. They were also advised that even though they had been told that they were being "toughened up" they were not yet tough enough to be fighting, so stay out of those, too.

Dave went to the airport located southwest of town and definitely outside the city limits. The date was July 18, 1959 and he rented a Piper Super Cub (PA-18) and flew for an hour with an instructor riding in the back seat. The most memorable thing about that flight was hearing an Eastern Airlines pilot call the control tower and request permission for a low pass followed by a climbing turn to the south and departure.

The controller approved the request from the Eastern Airlines training flight, which performed exactly as he had requested and Dave never forgot the beauty of the first Boeing 707 that he saw from the front seat of a Super Cub while on the downwind leg of the traffic pattern at Columbia as it performed the entry into a chandelle. (He had to describe it that way because it didn't change directions one hundred eighty degrees, only about eighty degrees.)

At the conclusion of the one-hour flight he climbed the stairs to the control tower cab where he spent the rest of the afternoon. Then he caught the City Transit Bus back to downtown Columbia, transferred to another bus and rode back to Fort Jackson. When he was asked what he had done on his weekend pass he said. "Just looked around and then came back to the Company area." He had made the resolution several months earlier to, "be a sweet child". One week later he had another "weekend" pass for twelve hours, so he took his first dual cross country from Columbia to Greenwood and return.

It must have been about the third weekend in August that Basic Training was completed and the troops were given "two week leaves". Dave's parents and brother drove from Evergreen to Fort Jackson in the '57 Plymouth, observed the, "Graduation Parade", visited the, "Company Area", met some of the cadre who were suddenly almost like total strangers to the troops. They behaved like nice people. The McKenzie family left Fort Jackson, drove to Augusta, Georgia and spent a night at Jewel's brother's home. The next day they drove to Evergreen where Dave spent the next fourteen days with nothing to do.

The Conecuh County Aero Club's Aeronca Defender had suffered another engine failure, but not while anyone was flying and it had been taken to a member's cow pasture and tied down beside a fence with its engine removed. It was caught by a strong gust of wind that pulled the tie down ropes to the wings loose and the airplane then tumbled over the fence backwards and came down inverted on the fence. It bent the fuselage structure and seriously damaged most of the wing ribs. The airplane that had been renamed the, "Offender" because it claimed to be an airplane was disassembled and taken to another club member's home where it was going to be rebuilt. Nothing was ever heard of it again.

At the end of the leave he returned to Fort Jackson and took his '57 Plymouth with him.

He was assigned to the, Wheeled Vehicle Mechanic's Course, Military Occupational School Number 628.3 probably because the Warrant Officer assigned by the Army to the National Guard Unit in Evergreen had found that typing those letters and numbers into the blanks on the form was always accepted. The course lasted eight weeks and during the first week the instructors learned that Private McKenzie already knew something about wheeled vehicles and they obviously passed that information ahead to the next course instructors in the school. It seemed that he wasn't asked to sit in the classes for the rest of the course. Instead he was given various minor jobs to do, but he did have to take the examinations every Friday afternoon on the subject covered during the week. At the end of the eight-week course he was named the, "Outstanding Honor Graduate".

He was then transferred to, OJT, "On Job Training" in a Motor Pool Garage and for the next two weeks his job assignment was to wash wiping rags in mineral spirits and hang them on the fence behind the shop to dry. He then became a "Leader" on a team of four "mechanics" who performed Q-4 maintenance on M-38-A1 four-wheel drive, ¼ ton trucks (jeeps). Every other Saturday afternoon he'd drive to Augusta to spend the weekend with his Aunt Mary and Uncle Elmo whom he soon decided was his favorite from either side of the family.

He was released on December 12, 1959 to return home. He went through Chattanooga again and on the way, spent a couple of days with the Moyers, and then returned to Evergreen and Southern Coach.

XVIII

LIFE RESUMES

Dave's next flight from Milton was a dual cross country in a Cessna 172 to DeFuniak, Springs, Florida and return with two friends, Marilyn and Bobby Stuart in the rear seat sharing in the cost on January 16, 1960.

That's also the date that Chuck Prince endorsed his logbook making it legal for him to make solo Cross-country flights and also to fly the Champion 7-FC "Tri-Traveler" solo.

A Certified Flight Instructor must endorse a Student Pilot Certificate for the holder to exercise the privileges of that certificate every ninety days. James R. McCutchan made the next endorsement on Dave's after riding with him for one hour in a Champion 7-FC. The 7-FC is essentially a Champion 7-EC modified to be on tri-cycle landing gear. It has one peculiar characteristic. It seems to need more weight on the nose gear. It's probably because the main gear is located too far forward on the fuselage. It is recommended that it be taxied with the control stick pushed into the nose down position instead of the reverse. That's probably the reason so many of them were tipped onto their back when attempting to turn from taxiing downwind to crosswind, a characteristic it shared with the Piper PA-22 Tri-Pacer except the pilot didn't hold the yoke forward most of the time.

Colonel James R. "Jim" McCutchan, USAF was one of those quiet, competent, capable, and confidant individuals that aviation seems to attract in great numbers. He loved to fly and at the time of this composition still does. Around 1940 he was a fighter pilot in the

U.S. Marine Corps and when Claire L. Chenault, who had been a Captain in the U.S. Army Air Corps. toured the United States' Air Bases seeking flying Officers who might retire from the U.S. Services James McCutchan was one of them that joined the man later known as General Chenault and flew in the American Volunteer Group in China known as the, "Flying Tigers". He was still flying in China when the A.V.G. was inducted into the U.S.Army Air Corps and flew the P-40, P-47, P-51, and P-38s in combat in the Pacific Theater until WW-II ended in 1945. When he endorsed Dave's Student Certificate he was stationed at Eglin Air Force Base and flew B-52's during the week. On weekends he'd come over to Milton "T" and fly any and everything.

Dave enjoyed meeting the people that he found on airports and around airplanes. He considered them to be deserving of recognition and respect and was very happy to be allowed the association with them. The confidence in his potential that was demonstrated by the endorsement of Colonel J. R. McCutchan, CFI-120284 is another one he is very pleased to have in his logbook.

Obviously he didn't enjoy the benefits of much support for his pursuit for happiness, confidence, and self-assurance. It appeared to him that his objectives were his alone and he had never enjoyed encouragement in selecting such activity. He always felt that the perceived sacrifices were well worth the benefits of his choices. The Prince family, that is father and son, had asked Dave in September why he didn't go ahead and complete the requirements for the Private License and he had to explain that he did not have the financial capability to pay for it at that time. Pete Prince, the father of the instructor, Chuck suggested that Dave concentrate on completing the training and then pay them when he could. Obviously, Dave accepted that offer and has always felt that he had friends in Milton, Florida.

Incidentally, Pete Prince was a Canadian that had moved from his birthplace and home in Canada to Springfield, Massachusetts as a young man and was employed by the Granville Brothers at their factory which produced the famous and successful "Gee Bee" racing planes during the early nineteen thirties. Pete Prince was the craftsman that built the wooden wings on all thirteen of the "Gee Bee" racers. On the very rare occasions when a person interested in aviation history could get Pete to sit down and begin talking about what he had seen and known the listener was treated to an extremely interesting history lesson, but real comprehension of the dissertation required that the listener had some basic understanding of the subject first.

On October 24, 1960 he flew the 1958 Cessna 172-B from Milton to Panama City, Florida to take the flight test for the Private Pilot License with the Single Engine Land rating on it from Designated Examiner, W. R. Sowell, Sr.. Mac rode from Milton to Panama City with him and when they taxied up to the Sowell Aviation hangar there was a band in full uniform with its member's instruments waiting for something on the ramp. Dave parked the airplane and his father and he got out, entered the office, and Dave introduced himself as Mac went to a chair in the corner and sat down.

Mr. Sowell collected his documentation and told him to plan a flight to some destination that has long since been forgotten. That was part of the oral exam that preceded the actual flight test and after reviewing the flight planning, the airplane performance calculations for the proposed flight, the weather and NOTAM checks, the Registration and the Air Worthiness Certificate, and the log books for the airplane and engine for the airplane Mr. Sowell simply talked about flying with Dave. Finally he asked, "Where's the airplane?"

Dave pointed it out and the two of them walked out, performed the preflight inspection on it and began the cross-country flight that had been planned. Mr. Sowell did, "pull the power", that is he closed the throttle simulating an engine failure after they completed the air work and when they were about four miles north east of the airport at Panama City.

Dave ran a quick cockpit check while turning towards the airport and said, "I'm returning to the airport." Mr. Sowell asked, "Really? Are you sure you can make it?"

"Yes." said Dave.

Mr. Sowell started shaking his head as though he was thinking, "No way." As Dave continued gliding towards the airport Mr. Sowell then mumbled, "Yes, I guess it's alright. I would have tried it."

Dave then explained that he did not have enough altitude to fly around the airport, so he was electing to glide across the field and make an entry to the downwind leg of the pattern from the wrong side, but it wasn't normally the correct way to do it. He made it and when he was flaring for the touch down on the runway he flew into the ground and bounced, but recovered it neatly. Mr. Sowell was just slightly smiling and said, "Now show me a soft field takeoff." Dave did and Mr. Sowell asked for a soft field landing. Dave flew into the ground again. Mr. Sowell asked for a short field takeoff and a short field landing. Dave bounced again. Mr. Sowell then said let's tie it down.

When they both exited the airplane Dave said to Mr. Sowell, "I'm sorry I wasted your time. I don't know what's wrong, but you didn't see me fly like I can fly. It must be nervousness." Dave tied the airplane down as Mr. Sowell walked away.

Dave walked into the office as Mac stood up and approached him asking, "Well, how'd it go?"

Dave answered, "I flunked it" and saw Mr. Sowell sitting at the typewriter filling out a form. He really thought it was a report of the failure. Mac didn't say anything and both of them waited silently for Mr. Sowell to complete the typing and explain what was wrong.

Mr. Sowell stood, approached the counter, and then just stood there as Dave approached him.

Mr. Sowell was holding some paper in his left hand and was not smiling as he slid it across the counter top to Dave.

Dave retrieved the slip of paper and started trying to read it. In the background he noticed Mr. Sowell's right hand moving across the counter and realized he wanted to shake hands.

He gripped Dave's right hand and said, "Three hours ago you came in here as a student. You may know something now that you didn't know then, but it isn't much more. Now don't go home this afternoon and depart on a world girdling flight tomorrow morning. This piece of paper says that you may now carry passengers while continuing to learn to fly. Take it easy. Don't try to do everything this week. Save some for the future. "CONGRATULATIONS."

Then Mac asked Mr. Sowell, "Was that band here to welcome us?"

Mr. Sowell explained that the Governor was supposed to arrive that morning and the band was for him. Mac and Mr. Sowell started talking and soon learned that one of Mac's former students became one of Mr. Sowell's former instructors.

Mac and Dave flew back to Milton "T" Field and had a very pleasant visit with the Prince's, Chuck and his father Pete.

The Spirit's Journey

The Civil Air Patrol Squadron in Evergreen had been organized in October of '58 and was called the, "Evergreen Composite Squadron". It had a collection of Senior Members who were by definition at least twenty-one years old and a "Cadet Squadron" whose members were fourteen through twenty years of age. They were authorized to wear U.S. Air Force uniforms with a unique emblem sewn over the right breast pocket that identified the wearer as a member of the, "Civil Air Patrol, U.S. Air Force Auxiliary" and the uniforms were decorated with the insignia of rank and any decorations that the wearer had earned in any branch of the military services. The wearers were expected to maintain and wear the uniforms in the same conditions that any member of the armed services wore theirs. The objective was to look like a well organized, groomed, disciplined, and competent branch of the nation's armed forces even though it had no authority, no compensation, little recognition, and no personal profit motivation for existence.

At the establishment of the Squadron there were four officers, Ceylon Strong, 1st Lieutenant, Squadron Commander, Ivey T. Booker, Lee F. Smith and I. L. Huggins 2nd Lieutenants while all of the other members were given the rank of Chief Warrant Officer 4. The only member that had any military experience or training was Lee Smith who had been a Sergeant in the Marine Corp.

The duties of the group composing a squadron were to participate in search and rescue missions for downed aircraft, be available for use in civil disturbances, assist emergency relief organizations like the Red Cross, participate in national disaster relief, and be available as might be required for use by any benevolent organization if the Air Force authorized their use. The Civil Air Patrol had no authority of any type whatsoever.

Dave joined because the Conecuh County Aero Club had been organized at the same time the Evergreen Composite Squadron, CAP was organized and had a requirement listed in its By-Laws stating that its members had to be CAP members. It was also the nearest thing to a social life that he saw in Evergreen.

(For the sake of brevity consider, "CAP" and the, "Aero Club" to be designations specifically for the groups in Evergreen)

In the spring of 1959 the Officers of the two groups, and they happened to be the same persons decided to have an Air Show sponsored by the Aero Club. The Aero Club would assume all of the responsibility and risk. The profits from the endeavor would be contributed to the CAP and Sunday, May 31, 1959 was selected as the date for Evergreen's first aerial extravaganza.

There was an ambitious "duster" pilot in Selma, Alabama that was hired to perform the main act in the air show, "Rex" Wiseman who would perform aerobatics in a Stearman that was used as a crop duster the rest of the time. Parachute jumps were to be performed by a Sky Diving Club that operated at Milton "T" Field in Florida, and the rest of the show was an aerial dusting demonstration by Jennings F. Carter from Monroeville, Alabama, and static displays of aircraft that happened to fly in. The Milton School of Aviation sent two airplanes to the show to sell sight-seeing rides. Emergency equipment and crews were supplied by The City of Evergreen providing a fire truck and Sam Cope, the owner of Cope's Funeral Home providing a hearse to be used as an ambulance.

Dave stood at the gate and collected admission from the attending spectators.

The show was presented without any problems. "Rex" Wiseman was paid his fee and left the site. Jennings Carter donated his services and left with his younger brother riding in the agplane's hopper. The sky-divers were paid a small fee and left without any complaints and the two airplanes with pilots who had carried the sky divers left with the proceeds acquired from the passenger rides.

The Conecuh County Aero Club had collected approximately twenty five hundred dollars and after payment of the expenses there was about nine hundred dollars that found its way into the CAP treasury.

It was the first time that an air show had been presented in Evergreen, but it was not to be the last. Two weeks later Dave left to serve his six month long military obligation at Fort Jackson, South Carolina and found when he returned that the City of Evergreen had decided to allow the Evergreen Composite Squadron of the Civil Air Patrol to move into and take over the maintenance of the original National Guard Armory and Meeting Hall. after the National Guard had moved to its new Armory, Fort Dave Lewis at the south end of Magnolia Avenue. The original armory had been used for Saturday evening square dances for the general public until the CAP was allowed to move into it.

The management at Southern Coach & Body Co., Inc. knew that the bus designed by Ceylon Strong and Allard French in 1941 was obsolete and Southern Coach was not likely to sell another bus unless a new and modern design was available. The United States Air Force Academy needed six new busses and a request for bids was issued. It had become obvious when reading the specifications in the Engineering Department that the author of the specifications was sitting in the rear seat of a General Motors Coach when he or she wrote them. The specifications described the bus used by Greyhound Bus Lines and built by General Motors Corporation, Truck & Coach Division.

PART XIX

THE SOUTHERN COACH S-41-DM

The Management Team at Southern Coach & Body Co., Inc. decided to estimate the cost of developing a totally new bus and submit a bid for that amount. The tactic was to develop the new product without it costing the small company anything. To accomplish that objective the company had to win the contract, therefore it had to submit the lowest possible bid to assure the award of it. If it proved to be successful the company would be able to develop the new product at no cost and would once again be competitive in the bus market.

Southern Coach won the contract for the forty-one passenger bus, diesel powered, with air ride suspension, and of course modern styling. The company had to deliver the six busses in six months. The Engineering Department had a staff of four Design Engineers and one person to maintain the files, run prints, answer the telephone, etc. The management hired Ceylon Strong to make the concept drawings, which he did by working on one Saturday evening and all day the following Sunday.

Edsel Williams, Roy Kendrick, Dave McKenzie, Alton Dean, Al Anderson, and Dewey Coburn went to work. The Purchasing Manager, Darwin Minninger was indispensable to the project. He was able to acquire the glass windshields for the G.M.C. bus, the V-8 diesel engine from G.M., the steering gear and column from Saginaw Steering, a Spicer transmission, Dana front and rear axles, seats, wheels, and tires from other suppliers. At times it seemed that if the designer wanted a part he'd visit Mr. Minninger and ask for it before he started designing it.

Everyone in the company was impressed with the wooden hammer forms built by Mr. Eugene Patten to form the aluminum panels to support the G.M. windshield glasses. They were a work of art that even the Italian custom body craftsmen would have admired. The "roll-out" power section designed by Dave that allowed the engine, radiator, and transmission to be removed from the rear of the bus as a single unit for major service or repair, was a marvelous feature.

The prototype was completed in approximately four months and the other five busses were in various stages of completion. Of course the test drive of the first bus with the passenger load composed of the office staff, shop foremen, etc. revealed one major problem and a few minor ones.

Dave got the assignment of redesigning the suspension system to eliminate the extremely objectionable sway of the vehicle on its air ride suspension. It required about two weeks of serious redesign to raise the roll center by raising the air springs in the bus and still required the addition of an anti-sway bar on the rear suspension, and load levelers in the air suspension system.

Southern Coach & Body Co. delivered the six busses in nine months instead of the six months specified in the contract. There was a problem with the baggage compartment doors and the main entry door that had to be resolved after delivery, but it was an achievement that did not require any apologies. When they were delivered in 1962 the six busses would have appeared to be up to date and current in any bus station.

PART XX

CAP, AN L-16, AND TWO AIRSHOWS

Even though the CAP is a United States Air Force Auxiliary the financial support that it receives from the Air Force is minimal at best. The National Headquarters is on an Air Force Base even though it is occasionally transferred from one base to another. Each state is designated as a Wing and some of the surplus uniforms and accessory clothing items are made available to the various Wings. There was a station wagon, a truck, a Link trainer, a portable electric generator, and a bus that had been built at Southern Coach that were issued to the Evergreen Composite Squadron. Maintenance and operation of that equipment required some financial resources. That was the justification for promoting an annual air show.

June 19, 1960 was the date selected by the staff at the Evergreen Composite Squadron and the Conecuh County Aero Club for the second air show to make the public more aware of the existence of the organization and to acquire funding for its operation. Again the Conecuh County Aero Club was named as the sponsor, but it was done for the benefit of the CAP. The only meeting involving the Aero Club had been the meeting in the Courthouse in September, 1958, but to be a member of the Flying Club a person was required to be a member of the CAP, hence the Aero Club became the, "'scape goat" to assume responsibility whenever the CAP Squadron needed money or did anything that might lead to responsibility being assigned.

The billing for the 1960 Air Show listed the Century Series Aerial Fly-over of USAF jets, the Alabama Air National Guard Jets Fly-over, a Fire-Rescue Team demonstration, the U.S. Air Force Dog Sentry Team, aerial Aerobatics, and the two Chimpanzees that had made the sub-orbital flight aboard a rocket for NASA. All of those exhibits and demonstrations were the result of the efforts of the Wing Liaison Officer, Major Lewis who deserved the credit for obtaining the participation of those listed exhibits and performances.

Dave considered the most impressive act in the show to be a demonstration by Mac's long time friend, Pat Mulloy flying a Schweizer I-23 sailplane. His act began with a release from the tow-plane at 3500 feet altitude, entered a spin over the departure end of runway 28, recovered headed east, looped, turned around by performing a half Cuban eight, followed by an immelman, did a series of rolls, another half Cuban eight, crossed the airport at an altitude of twenty (?) feet, performed a wingover turn around, extended the spoilers on the wings, and landed. The crowd of spectators was awestricken, silent, and speechless, it was so quiet and yet, there was not a sound of air rushing across the surfaces of the sailplane during the entire performance. When Pat Mulloy exited the glider after one of his sons removed the tape from the joint around the canopy the crowd burst into applause. Pat and his sons loaded the glider on the trailer and returned to Laurel, Mississippi. Jennings Carter had towed Mr. Mulloy aloft with his Bird biplane sprayer. (It was not the same Bird that Mac had owned, but Carter had allowed Mac to fly it once.) All of the emergency equipment that had been used the prior year was again available from the same sources and J. F. Carter returned home with his younger brother riding in the sprayer's tank.

Jewel had noticed that a lot of time and attention was expended on the Cadets in the Evergreen Composite Squadron and decided that perhaps Jimmy would benefit from becoming a Cadet. Mac enthusiastically welcomed that thought, joined the CAP in late 1960 and was the only member in the squadron who held a pilot's license. Jimmy went to one meeting and never joined or participated. He probably thought it was to costly, to regulated, and there was more "brow beating" and harassment displayed than he cared to tolerate.

Now that the squadron had a pilot the Squadron Commander, Captain Lee Smith found that the Alabama Wing Headquarters in Birmingham would transfer an L-16 from a squadron in Marion to the squadron in Evergreen. Mac, Lee, and Dave were to drive to Marion to meet a Designated Aircraft Maintenance Inspector on a Sunday morning in February, 1961. The "DAMI" would inspect the airplane and if he found it airworthy would then issue a "ferry permit" for it to be flown to Evergreen's Middleton Field.

Apparently neither Lee nor Mac knew that a ferry permit did not allow the carrying of a passenger because both of them got in the airplane and Mac flew it to Evergreen where Mac and Dave then proceeded to make the minor repairs that would bring its condition up to the acceptable standards before taking it to Pete Prince in Milton, Florida for the inspection and renewal of the airworthiness certificate. That led to a lot of paper work within CAP. The Alabama Wing was not allowed to fly its equipment into the airspace of another Wing and vice-versa. Part of Jimmy's opinion was valid.

After a couple of weeks minor repair work on the airplane the Alabama Wing and Florida Wing of CAP concurred and Mac flew the airplane to Milton where Pete Prince wanted to see the structure of the elevator on which Mac and Dave had patched the covering. That meant that Pete had to recover the elevator after inspecting the structure and he then renewed the airworthiness certificate approximately a month after he had received the airplane. Mac retrieved it in April 1961 and the CAP Squadron in Evergreen experienced an increase in morale. In retrospect when the performance and capabilities of an L-16 are considered the CAP squadron wasn't really any better off.

The L-16 had an 85 horsepower Continental engine that consumed five and a half to six and a half gallons of fuel per hour and cruised about eighty three to eighty five miles per hour depending on the power setting used. Its fuel system capacity was thirteen gallons, so when the pilot deducted forty five minutes of fuel required for a reserve fuel supply that left the airplane with a range of one hour and thirty six minutes if it was flown "conservatively" or one hour twenty five minutes if it was flown at the "blazing" speed of eighty five miles per hour. Its crew capacity totaled two people, a pilot and an observer. It really wasn't an airplane that was suitable for search and rescue missions. If the pilot had attempted to perform an effective search he would have created his own emergency.

The U.S. Army's use of the airplane was as an artillery spotter or liaison airplane that was expected to be based in the field close to the troops in need of its services where it could, "jump up and take a quick look around and then drop back down and out of sight." quoting the Information Officer based at the Army School of Aviation at Fort Rucker, Alabama.

On April 19, 1961 Dave flew the L-16 with his father appraising his performance from the back seat. At the end of the flight Dave thought he had, "busted" the flight check, but Mac signed the documentation that made him a "Pilot" in the Civil Air Patrol. He had a total flying time of one hundred two hours and fifty-five minutes. The Evergreen Composite Squadron, Civil Air Patrol now had a Command Pilot and a Pilot.

The CAP would not allow anyone not a member of the organization to ride in any of its airplanes, so Dave was flying solo in the airplane and it didn't require much time for it to become rather boring. Without the approval or knowledge of anyone in CAP he would loop, snap roll, and spin the L-16. Between April 19, 1961 and October 8, 1961 he flew the L-16 thirty hours and fifteen minutes that included five hours and 45 minutes in one day participating in a Search Mission based at St. Elmo Airport near Mobile. Four hours per month is the amount of flying time required by the military services for a pilot to collect, "flight pay". In 1962 he flew the L-16 a total of two hours and twenty minutes and in May'63 he flew it four hours and ten minutes participating in an Alabama Wing exercise based in Tuscaloosa, Alabama. The total time that Dave flew the airplane while in the Evergreen Composite Squadron was thirty-six hours and forty-five minutes.

"Mac" McKenzie giving rides to CAP Cadets in the L-16 in 1961. (author's collection)

CAP Captain Dave McKenzie and the L-16 at Tuscaloosa, Alabama, May, 1961. (author's collection)

October 1, 1961 was the date designated by the Evergreen Composite Squadron, Civil Air Patrol and agreed to by the Officers of the Conecuh County Aero Club for the third air show to be sponsored by the Aero Club for the benefit of the CAP Squadron, so the public thought.

Acts in the show were to be performed by a team of four U.S. Air Force T-33 jets performing aerobatics in a fly over, parachute jumps by the Pensacola Sport Parachute Club, a crop spraying demonstration by "Bill" Ellis of Andalusia, Alabama, an aerobatic performance by Jack Hale of Montgomery, races between a hot-rod owned by Eddie Edeker of Evergreen being driven by Dave McKenzie, and a Cessna 172 airplane flown by "Chuck" Prince of Milton, Florida. There was also to be a "clown act" that was not announced to the public until the show was in progress.

During the parachute jumps Eddie Edeker and Robert Ellington stood beside each other and started looking up at the airplane carrying the jumpers. One said, "Oh, there he comes. See him? He's waiting a long time to open the 'chute." The two of them coordinated the pivoting of their necks so that it would appear that they were following the descent of the parachutist as he fell. Then the other one said, "Oh well, maybe they have another one." and they both simultaneously looked back up. Some of the other spectators left the field.

The announcer of the show explained that one of the CAP Cadets was going to be given a flying lesson in recognition of his exemplary performance for the past year and to demonstrate to the public how simple a modern airplane was to fly the student would be allowed to fly the airplane solo (alone) at the end of the first lesson. The Cadet named was Barry Harper, son of Mr. and Mrs. Coy Harper of Evergreen. The instructor would be "Chuck" Prince of Milton and the announcer was Dave McKenzie.

The Cadet and "Chuck" Prince got into the airplane, taxied out, performed the pre-takeoff checks, took off on runway 10 and then landed on runway 18 as the announcer said that a student should have dual instruction in a cross wind landing prior to solo. The airplane rolled out of sight from the spectators, stopped, and the Cadet, Barry Harper got out, removed his CAP uniform shirt and gave it to a professional pilot, Glois Brand from Navarre, Florida who had been staged at the end of runway 18. After Mr. Brand got into the airplane it was then taxied back to runway 10 where "Chuck" Prince got out while the speaker broadcast the announcement that Barry Harper was now going to make his first solo flight. Mr. Brand took off and then went through the, "attempted landing", bounced a few times, added power and climbed a couple of hundred feet, turned around, dodged a tree, bounced a few more times, and finally landed the airplane. All of the airplane's gyrations were accompanied by sounds of panic from the announcer and someone was attempting to calm Coy and Mrs. Harper as the airplane taxied back to the parking area.

Mr. Brand got out of the airplane to very little applause for his performance. The majority of the spectators had believed what they had seen and heard was real and true.

It was Dave's first air show performance and the last for many years.

Cessna 172 flown by Chuck Prince & the Hot-Rod owned by Eddie Edeker & driven by Dave McKenzie in the air show at Middleton Field on October 1, 1961. (author's collection)

PART XXI

N-71568 ARRIVES

Dave had completed the payments on his '57 Plymouth on June 1, 1961, but the salary at Southern Coach was still only two hundred forty dollars per month and on Saturday, October 14, 1961 he decided that he'd drive down to Milton and join the Sky Diving Club. There were two sky diving schools in the United States that he knew of at that time. One was in Orange, Massachusetts and the other was in Milton, Florida.

He entered the office and asked Chuck who to speak to about skydiving. Chuck's response was to request the reason for asking and Dave answered that he was just tired of doing nothing and being bored. Chuck asked, "Didn't you say that you liked an airplane called a Luscombe?"

Dave asked why and Chuck said, "Go out on the line and look that one over."

Without knowing why he should look at the white Luscombe tied down at the north end of the line he walked out to it and feeling that he was authorized to look closely at the airplane he gave it the equivalent of a pre-flight inspection and returned to the office.

"Why'd you want me to look at it?" he asked Chuck.

Chuck asked, "What would you pay for it?

Dave explained that he'd have to talk to his banker at The Conecuh County Bank about a loan and see if he could withdraw his, savings from the office's Christmas Club at the other bank in Evergreen, and try to determine what he could pay.

Chuck asked, "When can you give me an answer?"

"Probably Wednesday afternoon." He answered.

That was the closest Dave ever came to jumping out of an airplane. He called Chuck on Wednesday afternoon October 18, 1961 and offered $1250.00 for the airplane.

Chuck said he expected that offer would acquire the airplane and the two of them agreed that Dave would come down on Saturday 21, October to pick it up. Chuck would call him if that offer were not agreeable to the seller.

On October 21, 1961 his logbook gained another entry from Col. James R. McCutchan that reads, "Luscombe check out O.K."

Most of September and October, 1961 David had been providing fuel and oil service at Middleton Field and scheduling students from Monroeville and Evergreen for The Milton School of Aviation and Jewel thought that Dave's request for Jimmy to go to Milton and drive his car back to Evergreen was so that Dave could fly the, Tri-Traveler back for the students to fly on Sunday. That was the little white lie that he told to satisfy Jewel's curiosity.

About 5:30 that afternoon Jewel and Mac were preparing to meet friends to go out for Saturday evening and while standing beside their car waiting for Mac to exit the house Jewel asked Dave, "Are they going to take that airplane you brought up here today back to Milton tomorrow?"

Dave responded, "I hope not."

As she began to frown and her eyes narrowed Jewel asked loudly and firmly, "Did you buy a damned airplane?"

Dave got in his car and backed out of the driveway. On Sunday morning he sneaked out of the house before his parents got out of bed.

About 2:00 o'clock Sunday afternoon he saw his parents' car approaching the paved parking ramp at Middleton Field where several people had gathered looking at Dave's new toy and everyone recognized Mac's car and became very quiet as their facial expressions began to show anticipation for the angry outburst that was imminent.

Jewel and Mac with friends, Trellis and Joe Shoemaker got out of the car and walked over to and around the airplane. Jewel simply frowned. Mac expended considerable effort attempting to hide any indications that he might approve of the airplane or the purchase. He did not want to break the silence. They got back in the car and departed as Dave loaded another passenger aboard the airplane and prepared for another flight.

The next time Jewel spoke to Dave was at breakfast the next Wednesday morning when she asked, "Did you really buy that damned airplane?" Before Dave could respond she added in a rather loud tone, "I OUGHT TO K-I-I-I-I-LL you."

By the time she had completed the statement Dave had gotten up from the table and headed for the bathroom to brush his teeth. She didn't speak to him again until the next Friday evening and that statement was almost civil. In the first seven days that he owned the Luscombe Dave flew it ten hours and ten minutes.

When Easter Sunday, 1962 arrived he was very comfortable in the Luscombe and had been introduced to a young lady named Mary whose home was located in the edge of the Conecuh Swamp south of Evergreen. Mary was a senior at the University of Alabama majoring in Education and seemed to get along very well with Dave. Dave got along well with her.

She had mailed a note to Dave advising that she'd be home for the Easter weekend and he responded by suggesting that they go to church on Easter Sunday morning. She accepted his suggestion and on Sunday morning he appeared at her parent's home in coat and tie wearing a boutonnière and presented a corsage for her to wear. They attended the services at the Evergreen Presbyterian Church and enjoyed lunch at Jimmy Water's Restaurant, the nearest thing to a good restaurant in town at the time and after the meal there was nothing to do in Evergreen, so Dave drove out to the airport and suggested that they fly down to Milton to watch the sky divers.

It was a warm spring afternoon and naturally there were some clouds and there were updrafts beneath the clouds. When there are updrafts there are also downdrafts. The result is that the airplane rides a little like a car on a rough road. The vertical oscillations in an airplane are longer in time and amplitude than normally encountered in a car so, the sensation may be new and seem upsetting to people that have never experienced it.

They arrived over Milton "T" Field and Dave pointed out the white "X" laid out on the runway intersection indicating that the field was closed and directed Mary's attention to the Cessna a couple of thousand feet higher than they were and said, "Watch the Cessna. There'll be some people falling out of it pretty soon." and started flying a holding pattern on the east side of the airport.

After about five minutes Mary grasped his right forearm and said, "D-D-D-ave, I- I- I f-f-f-f-eel s-s-s-sick."

He didn't usually carry passengers. There was a lot of opposition to aviation in the small town of Evergreen. Parents of young homosapiens of the feminine gender seemed to be especially opposed to their daughters flying with Dave McKenzie, so he had no need to carry the plastic product called, Sic-Sac's in his airplane.

Mary was beginning to make noises like, "Umph-umph-umph." and her breath had a slightly offensive odor, so Dave reached behind her and unlatched the door, pushed it open, released his hold with his left hand on the control stick, pushed the door open against the slip stream with his right hand, and with his left hand pressed Mary's head out the door. Their altitude was about two thousand feet above ground level.

To the southeast lay an airport named, Eglin 7 that was within and part of the Eglin Air Force Base reservation. It could be used for emergencies and Dave definitely considered his predicament to be an emergency and landed. After they had been on the ground a half hour Mary felt better, so they boarded the airplane and flew over to the airport at Milton.

As he taxied up to the hangar, office, and house Dave could see Mrs. Prince, Pete's wife with a huge smile watching them approach. When he turned the airplane and she saw the stains on the side of the fuselage her smile rolled to the inverted position. She took Mary in the house as Dave rinsed the airplane with the water hose beside the hangar.

After watching the sky diving for about forty-five minutes they got back in the airplane and returned to Evergreen.

He received one letter from Mary after that in which she asked, "Why don't you sell that airplane? All you ever say about it is to describe what part you're working on." He never answered her and never heard from her again.

Dave could recall his mother's opinion of airplanes and pilots and he was determined not to sustain the criticism and opposition to flying that he had observed Mac tolerating for so many years. If the girl didn't like airplanes she was not likely to be a friend of his.

PART XXII

A NEW PRODUCTION MANAGER JOINS SOUTHERN COACH

The first S-41-DM bus was complete, passed some testing successfully, and an unacceptable roll stability problem had revealed itself. Dave was working on that issue in the Engineering Department when a gentleman escorted by W. C. Bowers entered the room and was introduced as, Grady Thrasher, the new, "Production and Service Manager" of the company. The traditional courteous welcoming remarks were exchanged and office activity returned to its normal quiet routine as Dave wondered how the new management position had come into existence without any indication of such a position being made known to the faithful, long term members of the organization.

A few evenings later Dave was having his evening meal at "Jimmy's" which had been known as, "The Murphy Club" with a friend who was a DJ at the local radio station, WBLO, a five thousand watt AM broadcast station that "came on the air" every morning at sunrise playing the, National Anthem and signed off every evening at sundown playing, Dixie. Dave's friend, Terry Hammond the DJ had begun a relationship with one of the waitresses in the restaurant. Dave and Terry Hammond were engaged in casual conversation when Grady Thrasher entered, chose a table, and sat down while Terry and Dave continued their conversation. Of course, if you couldn't tolerate an occasional comment about aviation you would not enjoy Dave's companionship very long and in a few moments Grady asked from his seat, "Do you guys fly?"

That "shattered the ice". Dave had found a friend in Evergreen.

On July 6 Grady visited the Engineering Department and asked Dave if he'd like to fly to Bristol, Virginia over the weekend. A bakery in Bristol had purchased some delivery trucks with Southern bodies on them and had discovered some problems on them The top management of the company did not want the, "Production Manager" to be out of his office during the week and Grady felt that the appearance of the "Service Manager" at a customer's location during the weekend would be good for the company's customer relations, therefore he felt justified in proposing the trip. He also thought that it would be a fine example of the reasons for Southern Coach to add an airplane to its inventory of equipment. Dave liked that idea and agreed to go with his airplane. The only problem was the V.P. in charge of Engineering did not want Dave to miss the Saturday morning in the office, so they agreed to depart on Saturday afternoon and guarantee a return in time for opening the office on Monday morning.

Grady and Dave left Evergreen on the afternoon of July 7 in the Luscombe and made the first landing and refueling in LaGrange, Georgia. They landed next at Athens, Tennessee and waited more than an hour to obtain fuel because the airport attendant was not on the site when they landed. After loosing an hour of available daylight they decided to fly on to Knoxville and spend the night. They landed at the airport, on an island in the Tennessee River and at that time it had grass runways. The staff had left for the day. Grady used the pay phone on a post to call a taxi and then they had to climb the gate at the end of the bridge providing vehicle access to the south side of the island and await the arrival of the taxi to take them downtown to a hotel.

During dinner Dave learned a lot about the, All Star Air Circus which was founded by Grady in 1946 as the result of a visit by a gypsy parachute jumper at his fixed base operation in Athens, Georgia and his subsequent joining Hawthorne Aviation in 1950 and serving as the, Executive Commandant of Fixed Wing Training at Fort Rucker, Alabama. When employment in that position ceased to exist he joined Southern Coach and Body Company.

The next morning they rode another taxi back to, Island Airport, climbed the fence and waited for an attendant to arrive. No attendant ever appeared and they decided to proceed to the next airport on the route to Bristol,

Virginia. The weather briefing from FAA Flight Service that they obtained before leaving the hotel had forecast marginal weather conditions northeast of Knoxville, so they discussed the situation and decided to continue.

They considered landing for fuel at Jefferson City, but it was only about sixty more miles to Johnson City and they agreed that would be the end of the trip, one way or the other. They encountered the inevitable scene of the hilltops disappearing in the clouds. Dave recommended, "making a one eighty" and returning to Jefferson City, but a hole in the clouds appeared above them and Grady recommended climbing through the hole and proceeding via dead reckoning across the ridge to the next valley where another hole was bound to appear below them, "I've had to do this lots of times." he said.

No hole appeared in the clouds below them, so they continued straight ahead until they did discover one that they spiraled down through and started following the highway in a tunnel formed by mountains on each side and clouds above them. They were headed southwest when they stopped the spiral, so Dave reversed course and suggested to Grady that if they saw a country store with a gas pump he recommended landing in the highway and purchasing some fuel. Grady responded, "It's your airplane."

It was, "bumpy" in the valley and Dave could hear the float for the fuel gauge in the tank behind their heads banging on the sheet metal bottom of the tank occasionally and ahead was a river with the highway they were following beside it. A huge mountain rose into the base of the clouds on the southeast side of the highway and on the north side of the mountain was a field of grass. Dave said to Grady, "If we get through this narrow pass WE ARE LANDING in that field."

Grady replied, "O.K.".

As they approached the field they saw the runway at the Johnson City, Tennessee airport.

As they taxied up to the fuel pumps Grady said, "Don't say anything. Let me do the talking."

They had the airplane fuel tank filled and paid the bill. Then they read the receipt that listed the name of the airport they were on and indicated that they had purchased twelve and four tenths gallons of fuel to fill the thirteen-gallon tank. Calculations indicate that they had eight minutes of fuel remaining in the tank when they shut the engine down at the pumps.

The Spirit's Journey

While Dave was visiting the facilities in the office Grady had met and become friendly with a guy who had been out all Saturday evening and was hesitant about returning home on Sunday morning. He agreed to drive them to downtown Bristol and remain with them during the inspection of the bakery trucks, then return them to the Johnson City Airport.

They departed Johnson City about 1:00 P.M. E.S.T, landed in Athens for fuel, continued to LaGrange for another refueling and arrived at Middleton Field in Evergreen around 6:30 P.M. C.S.T.. It had been a long trip in a 65 H.P. Luscombe 8-A and one that Dave would NEVER FORGET.

He thought it was the closest he had ever come to losing his life.

In 1947 Grady Thrasher had taken two Ercoupe airplanes, and essentially removed the right wing from one, removed the left wing from the other, parked them side by side, and bolted them together creating the only four passenger, twin-engine Ercoupe ever built. He used it in the, "All Star Air Circus" from the time of its completion until the organization ceased to exist.

Grady Thrasher was a graduate of Georgia Tech University and really was a clever, industrious, and socially sophisticated person that Dave was very pleased to be able to claim as a friend.

The opening act in the Thrasher Brother's Air Circus with Grady performing in the "Twin Ercoupe" circling his brother, Bud making the "American Flag Parachute Jump" (photo from a postcard given to the author by Grady Thrasher)

Grady Thrasher was obviously a thinker and sometimes a dreamer whose mind would seldom rest. Soon after the flight to Bristol he proposed to Dave that they combine efforts, design, and build a "kite" to be towed behind a boat and be capable of carrying one occupant with altitude being controlled by the speed of the boat. The bank angle was not to be controlled, but lateral stability was to be maintained by the pendulum effect established by having the center of gravity located below the center of lift of the wing. A moveable rudder operated by the pilot's feet would influence the lateral position of the "kite" behind the boat towing it.

Grady pointed out that the kite did not have to meet FAA requirements so long as contact was maintained with the surface of the planet and it was not required that the pilot of such a device be certified. In other words, "It could not be flown in free flight." It would only be a toy for users to tow behind boats on lakes and rivers in the same manner as water ski's and inner tubes.

Dave advised Grady that he did not approve of the concept of having only one axis of control available to the pilot on the prototype, especially since both men involved in the development of the device were pilots and had some comprehension of the effect of flight control surface manipulation. Since Grady was financially supporting the project and providing the manufacturing facility, the garage behind his home. Dave acceded to his demands and suggestions. Both of them were concerned about the selection of an airfoil for designing the wing which was to be constructed by placing two aluminum spars in a concrete mold and pouring two chemicals into the mold. The two chemicals would mix, react with each other, and create a foam which would fill the concrete mold, capture the spars, and be trimmed before removal from the mold. After removal from the mold the wing would then be covered with fiberglass. The empennage surfaces were to be constructed in the same manner and the fuselage would be constructed of square aluminum tubing.

They made an aluminum template of the profile of the airfoil of an Aeronca, Model 7-AC "Champion" which Grady had borrowed from an operator he knew at Ozark to teach two students, Connor Warren and Lee Smith to fly at Evergreen. Dave began the drawings of the device while Grady began construction of the fuselage frame inspired by photos of the Bensen B8-M gyrocopter.

Dave wasn't aware that Grady was frustrated, felt that his efforts at Southern Coach were suppressed, and not appreciated by other members of the management staff. As a consequence, Grady began a search for a new employer, which led to his announcement that he was leaving Evergreen and assuming duties at Redstone Arsenal, the Marshall Space Flight Center in Huntsville. Mildred and Grady left Evergreen and took the "kite" project with them during the summer of '62

The next meeting of Grady and Dave occurred on June 27, 1969 in Huntsville where Dave had flown a Cessna 172 to visit his life long friend, Jan Hendrix and they visited the home of the Thrasher's. Questions about the glider-kite caused quite a display of amusement by Mildred Thrasher.

Dave's statement that he had responded to an advertisement for a brochure describing the device and had never received a response inspired a noticeable silence from Grady. Mildred Thrasher described how the completed device was hauled by trailer from their home to Lake Guntersville southeast of Huntsville for its test flight where it was unloaded onto the beach, assembled, and towed out into the lake by the boat. Then they pulled it close to the side of the boat, attached the tow rope to be used for the test flight, and Grady stepped from the boat onto the glider-kite's single foam filled fiberglass covered float with the seat, wings, and empennage all held in their appropriate relative positions by the fuselage frame made from aluminum tubing.

When Grady transferred his weight from the boat to the glider-kite he discovered that the assistant he had associated himself with in Huntsville had failed to correctly calculate the volume or displacement of the float required to support the glider-kite and its pilot. With the addition of Grady's weight it immediately sank until the wings were floating in the water. Grady was beneath the surface of the water and he had never learned to swim.

After rescuing Grady the "crew" towed the glider-kite back to the shore. Even though the float was made of foam covered with fiberglass and could not have absorbed any water the device had not risen after Grady's rescue. The wings kept it from sinking as they towed it back to shore, then onto the beach where they disassembled it, loaded the parts onto the trailer, and hauled them back to the Thrasher's home. They then stored all of the components in the garage. Mildred Thrasher asked Dave if he'd like to see it.

Grady broke his silence to state that the components of the glider-kite were buried behind and beneath a lot of miscellaneous junk in the garage and was not readily accessible for inspection. Mildred resisted challenging Grady's comments, but: her hint at a smile really said a great deal more than she could have verbally expressed at that moment.

David did not hear of the concept of, glass over foam aircraft construction again until six or seven years after his involvement in the glider-kite project with Grady Thrasher when Ken Rand in California introduced the Rand KR-1 monoplane, which used a similar technique for its construction.

The last time that Grady and Dave met was in May 1990 when Dave visited the Thrasher home in Grant, Alabama after Grady's retirement. Grady was recovering from a classic slip and fall accident that had occurred in his driveway during one of the rare snowfalls in Alabama.

It's easy to laugh at other's errors or mistakes, but it should always be remembered that the only people who ever made an error and were caught in it were those who did something.

PART XXIII

IT'S TIME TO MOVE UP

On Halloween night, 1962 the same young lady who had introduced him to Mary introduced him to Sherry Kirkland. Sherry was a member of a Baptist Church near her home twelve miles south of Evergreen and she seemed to believe that airplanes were not objectionable machines to have around. She didn't object to pilots either, so Sherry and Dave started seeing each other about once per week.

Dave started thinking about a move up to the Commercial Pilot Certificate and on December 22 he began to fill in the necessary qualification requirements in his logbook by taking thirty minutes of night-time dual instruction from "Chuck" Prince in a Tri-Traveler. That was the same night that he flew his first night solo. He already had two hours and fifteen minutes of night flying time with R. C. Crenshaw of Greenville, Alabama in a new Piper PA-28-160 Cherokee. Incidentally, that was the first year that Piper produced the Cherokee.

On December 30, '62 he had his first problem at night in the Tri-Traveler when he took Sherry to dinner in the terminal restaurant at Pensacola. He knew that Mr. Brand flew the airplane regularly on forest fire patrol, but he didn't know there was a radio under the seat used for communication with the State of Florida, Forestry Department that was not wired into the airplane in a manner that allowed the Master Switch to shut off the electrical power supplying it.

When Sherry and he boarded the airplane after dinner it wouldn't start. He had to teach Sherry how to handle the electrical switches, the fuel mixture control, and the throttle so that he could hand prop the airplane to start it. That is, he turned the propeller by hand. Of course, he left the tail of the plane tied down so that it couldn't get away from him if Sherry made an error. It started, Sherry did everything correctly and there was not a problem. After the generator brought the battery up they had lights and a radio, so he got a taxi clearance from the ground controller and started out to the runway and he noticed another airplane taxiing in front of him, but he hadn't heard anything on the radio.

He followed the other airplane to the run-up area at the take off end of the runway and while he was performing his pre-take-off checks he heard the other airplane's pilot saying, "Yeah, here it is behind my left wing. He has lights." The ground controller called, "87-echo, how do you hear?"

Dave answered, "Five by five."

"Call the tower when you're ready to go."

"Affirmative."

Dave knew then that the electrical load of the lights and the radio had drained the battery while they were taxiing and during the run-up it had recharged enough to bring every thing back up. He changed to the tower frequency, got a clearance for take-off with a right turn out towards Milton, acknowledged, and took-off. Then the tower said, "frequency change approved, 87-echo, good evening and don't bring that airplane back here."

"Roger, good evening", Dave responded and on December 31 he had his first two hours and thirty minutes of instrument dual instruction with Glois Brand.

In the summer and fall of 1961 when he was handling the supply of fuel for flight instruction in Evergreen Mr. Brand was the instructor and Dave was very impressed with what he saw and heard from him. In one of the group session lectures Mr. Brand was speaking about, "who's the best pilot and which pilot knows the most".

He said, "It doesn't matter who you are or how much flying time you may have you will always find a six or seven hour student pilot that knows something you never heard of or perhaps you heard and forgot. We don't know how much there is to know about flying, so when all of us combine our knowledge we may have ninety or ninety five percent of what there is to know accumulated in the total of us all or maybe all of us combined only know fifty percent of what there is to know. We have no idea so don't belittle another pilot. He'll always know something that you don't."

Dave wished Jewel could have heard that.

Mr. Brand's license number was 210074 and it is another one that he was always proud to have on endorsements in his logbook.

Dave took some more dual from "Chuck" Prince and Mr. Brand and on January 28, 1963 he passed his Commercial Pilot oral exam and flight test with FAA Inspector, R. L. Arendell at the Birmingham Flight Standards District Office. Immediately following that flight test Mr. Arendell had to get back in the airplane and fly without Dave wearing his glasses. He had the Statement of Demonstrated Ability for the Third Class Medical Certificate that's required for a Private Pilot, but he had to pass the test for the "SODA" applying to the Second Class Medical required by a Commercial Pilot.

As they got back in the airplane Dave told Mr. Arendell to double check the adjustment for the seat he was sitting in. The airplane was used to haul skydivers on weekends and the seat adjusting mechanism was not dependable. On climb out after the take-off Dave's seat suddenly slid all the way back on its tracks. His legs weren't long enough to keep his feet on the rudder pedals and he had to lean forward to hold the yoke in the correct position to maintain the airplane's pitch. He maintained control of the airplane and corrected the seat problem and while doing so he glanced over at Mr. Arendell who was attempting to cover his laughing at the situation. In later years he wondered if Mr. Arendell had pulled his seat adjustment lever to test his reaction. Stuff happens.

Mr. Arendell completed the forms for the SODA in the office and darkness was approaching as Dave departed Birmingham, Alabama headed south with Evergreen as his destination. He was alone, of course. There was no wind, the sky was clear, it was cold and it was the first time in his life that Dave had ever sat in one seat and noticed the beauty, the pleasure, the feelings of security and contentment that the observation of a sunset can provide. That night he flew back to Evergreen, picked up Mac and got some friends to drive to Milton to bring them home. He was then a Commercial Pilot with three hundred hours and fifty minutes total time. He had achieved a goal that his school teachers, classmates, friends, and mother had declared to be unobtainable. He

COULD fly. He had flown. He was now a commercial pilot. Now what would he do with the certificate, the skill, and the knowledge that it represented?

At three hundred thirty seven hours and forty five minutes total personal flying time he sold the Luscombe N-71568, spent the money to purchase appliances and furniture, rented a small house on Salter Street in Evergreen and on November 2, 1963 married Sherry Kirkland.

PART XXIV

THE CLARK CORTEZ LEADS TO A FUTURE

In late 1962 Southern Coach's management bid on production of a small motor home called, "Cortez". It was an advanced design motor home that had been the brainchild of an executive at Clark Equipment Company in Battle Creek, Michigan who was fascinated with motor homes. It was small enough to park in a regular parking space suitable for an American automobile and yet would sleep six adults, had cooking and eating facilities, a toilet, and a shower. Its design featured independent torsion bar front suspension and trailing arm independent rear suspension, front wheel drive, power steering and brakes, air conditioning, natural gas heating, and a five speed manual transmission. Southern Coach won the contract over most of the body builders in the United States and work began on it in Evergreen.

The Junior Van had filled the market's demand in late '61 with approximately three thousand five hundred vans delivered and that was the last year that the Elrod plant was used to produce it. The Cortez was an ideal replacement and the first one hundred units were completed and delivered by September 1963.

Dave had become the engineer who worked primarily on the Cortez and Clark Equipment asked for one hundred twenty four minor changes called, "model changes" to be made on the design before the start of the 1964 model year's production began in early '64. When Dave advised that such a demand was unrealistic and impossible for one man to deliver in the three months remaining before production of the '64 model was scheduled to begin Clark Equipment hired five designers from Modern Engineering Service Company in Berkley, Michigan to fly down to Evergreen and perform the requested minor design changes on the vehicle.

Dave was then instructed to establish a drafting room in one of the old buildings at the former WW-II vintage venetian blind plant. Approximately September 15, 1963 the five designers from Modern Engineering arrived, walked around the room looking at the layout, selected working stations, and moved in. In less than a week everyone working on the project was on a very friendly and first name only basis. The informality in the work place was new to them and they loved it. Jokes about Southerners and Alabama flowed freely. They asked many questions about life in L.A., that's Lower Alabama and were unbelievably surprised at the friendliness of everyone they encountered. The design changes were being made very quickly and smoothly. The morale was the best that Dave had ever seen in an engineering office.

The guys from Modern couldn't find a place in Evergreen to live in, so they rented a small house on the north side of Andalusia that was the basis of one of the jokes. The wife of one of the guys was expecting delivery of a baby at any time and the rest of the guy's in the office started teasing him about naming the baby, Andalusia if it was a daughter. He took it well and advised them that his wife was looking forward to seeing them again.

Alligators was the subject of conversation one day, so the room next door to the office was the plant's warehouse where an African American stock handler worked. He was called into the office and advised that they needed a lizard to mail home to one of the guy's son. Willy promised to provide one that he'd catch in the woodpile across the street. When they all returned from lunch they each had a fly swatter and for the next three or four days a vigorous "swat" was occasionally heard followed by scratching noises. They were accumulating dead flies to put in a box with the lizard to eat if one was ever delivered and they'd mail it to the youngster.

It all came to pass. The lizard was placed in a cardboard box that had small air holes punched in it and the carcasses of the flies were placed in the box as a food supply. Then it was taken to the post office on Rural Street in Evergreen and mailed to the youngster in Livonia, Michigan. The youngster took the lizard to his biology class in one of the Livonia High Schools and since he believed it was a baby alligator he had no difficulty in passing it off as a baby alligator even to the biology teacher. He was in trouble with his mother though when he lost it at home.

In the third week of December the four remaining guys that were still working on the project completed it and returned to Modern Engineering in Berkley, Michigan.

Photo of the "Cortez" motor home copied from the sales brochure. (author's collection)

PART XXV

THE CAREER PATH LENGTHENS

Friday afternoon January 3, 1964 a telegram from Carl Herbart, the Team Leader at Modern Engineering who led the designers making the design changes for the '64 Cortez arrived asking Dave to call him Saturday morning.

Dave thought it was probably concerning another engineering issue on the Cortez project, so he responded by telephone and was extremely pleased, complimented, and surprised when Carl Herbart explained that the telegram was to open negotiations for employment of Dave in Michigan and he offered a choice of projects for Dave to consider. The first choice was to work on projects for material handling equipment being designed for Clark Equipment Company and the second choice was to work at Ford Motor Company's location on the design of automobile chassis. Dave was both stunned and elated.

Carl and he agreed that he would make a decision within the next week and advise him as soon as it was made. In a period of one year he had made the major steps in achieving both goals that the six year old boy had chosen that day in Mrs. Weather's first grade classroom when the teacher and others had said, "You can't fly." and forced him to select an alternative goal. He had said, "Well, I want to work for Mr. Ford.". Here was the alternative goal coming into view.

Later that afternoon he telephoned Mr. W. C. Bowers, the Vice President in charge of Engineering at Southern Coach to make arrangements for Sherry and he to visit him at his home early that evening. Dave opened the conversation by pointing out that since beginning a family he needed to revise his priorities. He needed to place more importance on personal financial growth and told Mr. Bowers about the offer from Modern Engineering. He did not reveal the expected amount of compensation which was twice the amount being paid by Southern Coach. Mr. Bowers asked to be advised as soon as possible what decision Dave made.

Monday morning he told Mr. Bowers that he would like to have a salary increase to five hundred dollars per month. It was still a lower amount than offered by Modern Engineering. As expected Mr. Bowers objected very strongly to the request and they ended the conversation with the understanding that Mr. Bowers would contact him with a counter offer on Tuesday morning.

When Mr. Bowers called him Tuesday morning and advised that the request for five hundred dollars per month was absurd and ridiculous Dave had to respond with the announcement that he would accept the offer from Modern Engineering and suggested that they agree on the separation date from Southern Coach. Dave attempted to explain that he personally felt that he owed the company the time to plan a transition for the changes that would obviously be necessary.

Mr. Bowers advised him that, "I don't give a damn if you pack up and walk out right now. I don't even give a damn if you pack up and I don't care whether you walk or run."

The response was, "Go." and Dave said, "O.K., Friday afternoon January 10". Neither of the two men was pleased about the outcome of the conversation.

Dave had to contact his landlord and tell him of the changes and agree to continue to rent the house to store their furniture until they found a place to live in Michigan and on Sunday morning January 12, 1964 Sherry and he were in the '57 Plymouth with all of their personal items such as clothing in the trunk and rear seat driving northwards on Interstate 65 enroute to Detroit, Michigan and an unknown future.

They drove through the mid-west's worst snowstorm in the winter of '63-'64. Their first encounter with the snow occurred while crossing the Alabama-Tennessee State Line at about 7:00 P. M. and it required three more hours to reach Nashville where they stopped at a motel for the night. The next morning they resumed the journey going through

Louisville, Kentucky and reached Cincinnati, Ohio at 6:00 P.M.. It required eight more hours to reach Allen Park, Michigan, a suburb on the south side of Detroit where they spent the remainder of the evening in the Allen Park Motor Lodge on the first night that it was open for business.

On Tuesday, 14 January they called Carl Herbart at Modern Engineering and advised him of their arrival in the area. They then moved into the Drake Motel on Woodward Avenue in Royal Oak on the north side of Detroit and waited anxiously for Dave to report on Wednesday morning, January 15, 1964 for his job assignment at Modern Engineering Service Company in Berkley, Michigan.

At 8:00 A.M. Dave reported to Modern Engineering and completed all of the forms required by an employer for a new hire and was given directions to Building 5 at Ford Motor Company's Research and Engineering Center in Dearborn. He was given a telephone number and told to enter the lobby on the east end of the building and to call Mr. John Caldwell.

Mr. Caldwell entered the lobby, introduced himself and said follow me. They rode an escalator to the second floor and walked the length of the building to a room completely full of drawing boards like none that Dave had ever imagined. They were five feet wide and twenty feet long covered with a sheet of aluminum anodized an off-white shade with a grid of lines spaced ten inches apart running horizontally and vertically. At each board was a gentleman in a white shirt wearing a tie. Some were standing, some were seated, and some were even lying on their drawing board in order to reach the area of the drawing they were making. John Caldwell led him to a desk and said to the man sitting there, "Paul, here's the new man from Modern." Then he walked away.

Paul Ceru led him to the north side of the room and introduced him to Rod Pharis whom he was going to share the board with. He then started unloading his attaché case and was introduced to most of the guys around him. Of course he knew it would require three or four days for him to become comfortable, but he was starting and in a short time another gentleman walked up and said here's a job for you. We have this new wheel cover proposal and need to conduct a curb clearance study on it.

That evening he returned to the Drake Motel in Royal Oak, picked Sherry up and they went out to dinner. They had rented a "kitchenette" for a room and had purchased a few groceries on the afternoon of the fourteenth, but dinner had to be more than they had available in the room. That's how the rest of the week passed. Dave commuted to Dearborn where he parked his '57 Plymouth in the Executive Parking Lot of Ford Motor Company's Building 5 on the Scientific Research and Engineering Center Campus and finally learned where the employee's lot was located on Friday. He learned the names of a few of his fellow employees, but he also found that he had a language problem caused by his coworkers listening to him with their accents.

In south Alabama the family heritage is mainly English, Scotch, Irish, and a few French. He had never met descendants of the immigrants from the eastern European countries and he found that residents of the northern industrial cities have their own unique accents. They alleged that he had a very heavy accent and his workstation assignment placed him next to the telephone intended to serve approximately ten guys and mounted on a post in the area. The telephone would ring, he'd answer it and be asked, "May I speak to Joe Btpsflak?" or something like that and he'd have to ask the caller to spell the name. When he wrote it down he found most names spelled with several consonants and only one or two vowels. He'd still be unable to ascertain the pronunciation and would have to ask for assistance from someone else to locate the person that the caller was attempting to reach. After about two days he announced to his co-workers that he was not going to answer the telephone again until he learned the correct pronunciation of all the names. He wasn't being rude or snobbish. He was simply trying to avoid becoming an obstacle in the communication process. His co-workers smiled, smirked, looked at the ceiling, nodded their heads, and some said, "We understand." That was also the first time that he heard the name, "Polack".

On Friday evening he called the home of one of the guys who had been in Evergreen working on the Cortez, Stan Zylinski and Tina, Stan's wife. Stan, Sherry, and he agreed to attend the annual auto show at the National Guard Armory in Oak Park on Saturday evening. Dave had seen the photos and read the articles in the auto magazines about the Auto Shows in New York, Chicago, Los Angels, Geneva, Switzerland, Frankfurt, Germany, and other cities around the world, but this was to be the first opportunity in his life to actually attend an auto show. He left the show feeling slightly disappointed because with the exception of six or eight "concept" cars from the manufacturers he hadn't seen anything that he couldn't find in a dealer's show room.

On Sunday the Zylinski's asked Sherry and Dave to temporarily move into their home in Berkley and stay until they could establish one of their own. Stan also explained that he was working from Monday through Friday in another city and both he and Tina would feel better if the McKenzie's were in his home with Tina. Sherry and Dave moved in that afternoon. It's great to have friends and Dave wondered why there were so many jokes about, "Polacks".

On Sunday, February 2 all four of them drove out to a small airport in Fraser named McKinley, rented a Cessna 172 and went flying for about an hour. During the hour Dave noticed what appeared to be an automobile racetrack and later learned that he had seen the Packard Motor Co. test track. How exciting it was to him knowing what an historic area of the United States' automobile industry he was moving into. An understatement would have been to say, "He was elated." Tina believed that only the angels should have wings and she was already afraid before they left home. Tina did not like flying and never flew again.

Dave had decided that because he was a "Jobbie" or a "Gypsy" as the Ford Motor Company employees referred to temporary employees he had better locate a house to rent reasonably close to the home office of Modern Engineering, so Sherry and he had found a house to rent at 222 E. Lincoln in Madison Heights and on the third Friday in March Dave received a phone call in his office from Sherry advising him that the truck hauling their furniture had arrived. He advised his supervisor of the development and was released from the office to drive home and move in. They had been contributing financially to the Zylinski household, but they still felt that there was no way they could ever repay them for their generosity and assistance for the past two months.

Life in Michigan became as nearly normal as it can be for a commuting, "Jobbie" as is possible until he had an accident on the way to the office in May. Then while the Plymouth was being repaired he had to join in a car pool for transportation. He was picked up at 6:40 every morning at home by another Modern employee, driven to the office in Berkley where he waited for others in a car pool to arrive and then all of them rode to Dearborn. In the afternoon at 4:30 they left the office at Ford, drove to Modern's office and then rode home with the same person whom he left with that morning. That routine was followed until the end of June.

In April the '65 Mustang had been released to the automobile market in the United States and the car was an instant and unprecedented success. Dave was one of several designers in the, "Advanced Chassis Section" of Ford Motor Co. that was driven to the Dearborn Test Track to see the car, a red convertible. He fell in love with it and tried to buy one, but no Ford dealer would discuss a trade for a '57 Plymouth sedan and Ford Motor Co. would not authorize a discount called, "A" plan for a, "Jobbie" and by August 1 he really needed a new car, so he visited a Plymouth dealer in Royal Oak and purchased a new '64 Savoy two door sedan just weeks before release of the '65 model. That move certainly raised some eyebrows in the office at Ford.

By that time he was attempting to design an adjustable brake and accelerator pedal system to hopefully be introduced to the public in the '67 Thunderbird. The proposed accessory was never offered and Dave worked on other projects while hoping that Ford Motor Co. would respond to the application he had placed on file and offer employment to him. In the meantime Ford had gone through one of its periodic reorganizations and his supervisor, Paul Ceru had been transferred to another section and replaced by Bob Murphy. Mr. Murphy was promoted to Department Manager in the next reorganization.

On Friday afternoon, October 28, 1965 Dave was one of the few guys still in the room at 4:30 and he heard a familiar voice over him. He was lying on the board designing a part for a chassis frame when Bob Murphy asked, "Do you think you can work for me?"

"Well, I used to. You tell me."

"At 8:00 o'clock Monday morning go to medical in Ford Division Headquarters. They'll give you a physical and tell you where to go from there."

On Monday, November 1, 1965 the other goal that he had chosen at the age of six in Mrs. Weathers' first grade classroom became a reality. He started working for Mr. Ford.

PART XXVI

STARTING A HOME & LIFE GOES ON

The third person that Dave met when he entered Ford as a, "Jobbie" was another designer, Rod Pharis, whom he thought was probably the most productive and talented designer he ever knew. Rod had during his high school years developed an interest in "U" controlled model airplanes and had designed a mid-winged monoplane that used an engine whose displacement placed it in "Class C" of the Academy of Model Aeronautics competition rules. He had won the National "Stunt" Championship two times while still a teenager.

The acquaintance with Rod Pharis inspired him to revive his interest in model airplanes since his family responsibilities and financial resources were causing such radical changes in his life and Sherry had contacted Rod prior to Dave's twenty eighth birthday and received his assistance in purchasing a .35 cubic inch, "Super Tiger" engine as a gift.

Dave designed a biplane model, built it, and asked Rod to perform the test flight for him.

They met in Rouge Park on Detroit's west side one Saturday afternoon in the spring of 1965 and Rod flew it. He said afterwards that it was the most control sensitive model that he had ever flown. It required Dave to think about it for a couple of days and then he moved the wheels farther forward to shift the center of gravity forward to be even with the leading edge of the upper wing. To Rod's surprise Dave started flying it successfully. Dave "crunched" it on the tenth flight because he was attempting some aerobatics with it.

Rod then let him have a copy of the drawings of his "profile" midwing, combat class model called, "Banzai". Dave modified the design by adding tri-cycle landing gear to it which increased the weight of the model, but Dave chose not to reduce the wing loading by increasing the wing area. It flew quite satisfactorily and, with his strong interest in aerobatics Dave, "crunched" two of them in about six months.

Dave hadn't quit flying. On March 22, 1964 he had found a small airport that had one runway of gravel and a second one of grass named, "Big Beaver" not far from the house that Sherry and he were renting. A lady and her brother, Anna and John Main, owned it. After Anna inspected his logbook she allowed him to fly their two Aeronca airplanes without taking a check ride with an instructor. He thought that was a huge compliment from Anna. And on June 28 he checked out in the Tri-Pacer owned by Gil Baker and leased to Anna and John. He used the Tri-Pacer to take Sherry, Rod, and his wife, Sue out to a lake northwest of Detroit for the Pharis's to have an aerial view of a lot they had purchased.

Anna started calling on Dave occasionally to run out to Big Beaver and carry a load of passengers on a pleasure flight and there were a few flights made for a photographer to shoot aerial views for a customer. On September 12, 1964 he removed the rear door on the Tri-Pacer to allow the photographer to lean outside the airplane with his camera to take the photos on a trip from Big Beaver to Lansing, Ann Arbor, Ypsilanti, East Pointe, and return. The photographer asked Dave to climb as high as he could for the shots over Ann Arbor, Ypsilanti, and East Pointe. Dave was able to coax the airplane to 13,000 feet during that session. He may have been able to go higher, but was pleased to hear the photographer say, "This is high enough." Dave was getting cold, but he was earning money with his Commercial Pilot Certificate. He wasn't, "Taking it out of the family's finances." as Jewel used to say.

On August 30, 1964 Sherry gave birth to their son in Martin Place East, Hospital in Madison Heights, Michigan. Michael David McKenzie was robust at a birth weight of ten pounds, eleven ounces and both the mother and father were pleased to take him home. Sherry's mother had traveled to Michigan a month before his birth to be with her daughter during the first few weeks of his life and Dave's parents and brother drove up from south Alabama about two weeks later. "Gran'ma

Jool" as Mike later addressed her was concerned about a possible mix-up in babies in the big hospital. She later said that her first glimpse of the baby boy dispelled any consideration of a mistake or exchange.

During the lunch break one day in the fall season of 1964 a guy walked up to Dave's board and introduced himself as Rodney Beckwith, III. He claimed he had been wondering where the guy who drove the gray Plymouth was located in the building. He had seen the car on the way to work several times and wanted to discuss the possibility of organizing a car pool. He lived in an apartment in Royal Oak with his wife who was employed by an Insurance Agency in Detroit.

The two of them made an agreement and for the next six months Dave would drive from his home in Madison Heights to Rod Beckwith's apartment in Royal Oak and pick up Rod and his wife, Georgia every day for one week. The next week he'd leave his car at their apartment, get in their car, every day and ride to and from work. In early spring the Beckwiths moved to Inkster where they had purchased a three bedroom, brick ranch type house with a full basement. Of course that destroyed the, "car-pool".

In July, 1965 Rod told Dave that he had been talking to the neighbor who owned the empty house next door to theirs and offered to introduce his acquaintance from Ford. The McKenzie's rented the house identical to the Beckwith's home except the front door location and moved into it on August 15, 1965. In October 1966 Sherry and Dave became the owners of the house next door to the Beckwith's.

Dave had not felt well several times that summer and neither he nor Dr. Claude Oster could determine what the cause might be. Dave decided to "clean up" his life style. He'd quit watching the late movie on TV every Friday and Saturday evening. It helped, but not much. The reduction in the length of his work day helped when he moved to within five miles of the office instead of the twenty two miles that he had been commuting each way, but he still wasn't up to full speed. Perhaps there is some truth in the allegation that stress is not conducive to well being. After Ford hired him he had gained some employee benefits of which one was called, "sick leave". In late November '65 he came down with the flu and stayed home a full five days. He didn't think that it was "going to fly."

When he started college in June '55 he had never smoked, but in October '55 had allowed his playmates and co-workers at Kimberly-Clark Corporation in Memphis to introduce tobacco consumption or addiction as a stress reliever and "cool looking". He was hooked and in the fall of '65 he was smoking two and half packs of cigarettes per day. He weighed one hundred nineteen pounds and he didn't feel, "cool". In March '66 he quit smoking, "cold turkey", completely, and by will power only. The first week was awful, but he kept telling himself that he was a fool if he didn't stick to the decision to quit. By the end of the second week the urge to "light up" occurred only four or five times per day and the cure was to become mentally preoccupied with something, anything else. At the conclusion of the third week he knew that success was close at hand and by the end of the fourth week he knew that he HAD SUCCEEDED. During the next six months he'd feel a desire to smoke about once every week that slowly shrank to once every two weeks. He kept reminding himself that if he yielded once he'd be back at the beginning. At the end of six months he was free of that despicable dragon.

He could remember the time in December 1956 when Jan Hendrix was going to return to Tuscaloosa, Alabama and the University of Alabama. His father, Dr. Hendrix wanted Dave to go along on the trip and drive the car back to Evergreen. On the way from Evergreen to Tuscaloosa Dr. Hendrix stopped at the museum in the Moundville State Park and they toured it.

As best as has been determined those Indians had come northward from Central or South America across the Gulf of Mexico long before either of those geographical areas had names and had entered what was later known as, Mobile Bay, went up the Tombigbee River and settled in the area that became known as, Moundville. When the Spanish explorer, Hernando DeSoto led his expedition through the area in the 1540's they met the nearly white skinned Indians and described them as the, "Alabamos" or something very similar that meant Albinos, hence the name of the State, Alabama.

The museum was built around and over an Indian burial mound that had been opened by archaeologists or anthropologists and the second floor was actually a walkway around the open mound. Dr. Hendrix was the Conecuh County Coroner and Pathologist and began to point out symptoms that he could see and identify on the skeletal remains of some of the Indians.

He was saying, "See, that one was thirteen to fifteen years old and died of small pox. The other one over there was in his mid-twenties and died of old age. This one over here was about seventeen and died of syphilis." He pointed out several that died of the venereal disease. Then he said, "See, after the Spanish gave the Indians syphilis and Christianity they had to return the gift and gave the Europeans tobacco."

At Ford, Dave was working ten hours per day, Monday through Friday with eight hours on Saturday. There was a three week period in March '66 that he didn't have a day off for three weeks, but he enjoyed the job. He was working on the steering columns, transmission control linkages, brake, clutch, and accelerator controls, the parking brake system, and fuel tanks in the '67 Falcon, Galaxy, Mercury, Mustang, Montego, Continental, Lincoln Town Car, and Thunderbird during that spring. He felt pretty good. His design leaders had been changed every time his job assignment had been changed. He began to think that the management believed he could do anything. They probably thought of it as only drafting, but drafting is the designer and engineer's language that is used to tell the technicians and craftsmen what they wish to have or what they're thinking.

Perhaps the executives at Ford were uneasy about a Plymouth being parked in a Ford Company parking lot. Evidence that such was the case was delivered by the U.S. Mail carrier to 26034 Woodbine Dr. in the form of a note from a gentleman named Dave Evans that stated, "We know our competitors produce fine products, but we believe it expresses confidence in our organization if the employees drive our products."

Dave composed a note in response to Mr. Evans' letter stating that he completely agreed, but he was unable financially at that time to purchase a new vehicle. It definitely was his intention to make his next purchase a new Ford product. He did purchase a new 1970 Mustang fastback coupe in December 1969 and every automobile that he purchased afterwards was a Ford product. Yes, he did have confidence in them.

PART XXVII

THE DESIRE TO FLY IS WELL AND STRONG

In January 1966 he was walking past a "Jobbie's" station and noticed the copy of a magazine lying on the end of the board. He stopped, introduced himself to Bob Witzke, picked it up and it was the first time that he had ever seen a copy of the Experimental Aircraft Association's publication, "Sport Aviation". Bob suggested that he take it to his board and look through it at lunch and then BRING IT BACK.

Over the next few days Dave and he had a few conversations and they agreed to go to the next EAA Chapter meeting together at, Allen's Airport located on the east side of Pontiac, Michigan. At that meeting Dave knew no one except his new friend, Bob Witzke. He was still impressed with what he saw and heard that evening even though no one seemed to care whether he was there or not. They were totally engrossed in discussions about various problems they were having while attempting to construct components for their airplanes that they referred to as "projects". The meeting had opened with a five-minute report to the membership by the President on the decisions and actions taken by the officers. That report was followed by a presentation on how to perform some operation requiring mastery to build an airplane.

Dave later learned that one of the Chapter's by-laws specifically stated that the officers would run the Chapter and open each meeting with a report on their actions and activities that was not to exceed five minutes in length. It wasn't stated in the by-laws, but the members said, "If we don't like what they do we won't re-elect 'em." The officers didn't seem to be concerned about it though.

Dave went to a meeting of the oldest Chapter in the Detroit area at McKinley Airport and a meeting of the second oldest Chapter in the area at Mettetal Airport. He then elected to join Chapter 194 at Allen's Airport in February 1966.

He continued to work in the same department at Ford and the reorganizations had decreased in frequency. In the six months that he had been an employee he was working with his third supervisor. A gentleman named Roxxi Rossi that would hold the job for three more years. His section had a morale problem that Dave couldn't find any justification for. Yes, Roxxi always toured the room on Friday afternoon asking all of the designers if they could come in for five hours on Saturday morning. Everyone accepted it as standard practice, but Dave asked Roxxi why he couldn't ask on Thursday because he always submitted the request for authorization of overtime on Wednesday and had the authorization on Thursday. Why did he have to wait until Friday to ask the designers? Dave thought that if the designers were informed of the schedule on Thursday they'd have a little time to plan their weekend. The request was honored, but Dave never felt at ease about it and the job became a routine to him until the summer of 1969.

In 1966 he met another person interested in flying that was working in the same general area of the building and was a member of the Civil Air Patrol. Dave attended his first meeting with John Matusiak at the Crown & Sword Restaurant & Bar in Dearborn on a Wednesday evening. The group was called a Senior Pilot Squadron. It didn't have any equipment except an L-16 that was assigned to the Squadron when it was formed some years before. The L-16 needed a major rebuild and recovering when they had acquired it and they rebuilt it as a club project. When Dave saw it he could understand why it was almost, "out of annual". That is, not licensed or the Airworthiness Certificate had expired because it would not pass an inspection by a Designated Airworthiness & Maintenance Inspector. They appointed Dave to be the, "Maintenance Officer" and he found a DAMI to perform the annual inspection. The group just met at the Crown & Sword and drank beer.

During the inspection the DAMI found that when the guys in the squadron were recovering the wings they had left the fabric reinforcing tapes off the fabric over the ribs. There was also an Airworthiness Directive on the carburetor that had never been complied with.

Wayne Platner, the DAMI became distracted by other projects that he assigned a higher priority to and progress on completion of the L-16 seemed to stop. Of course, Dave was assigned blame for the lack of progress by other members of the squadron, but Dave remained with the group until the airplane was "relicensed" and one other member, Mel McGee and he were flying it.

Dave had attempted to renew his membership in the Civil Air Patrol and didn't receive any response. He asked the guy who was supposedly the Squadron Commander about it and he said the check and forms had gone to Michigan Wing Headquarters and they hadn't heard anything about it from National Headquarters.

This certainly was not the type organization that he belonged to in Alabama three years earlier where he had earned recognition as the Outstanding Information Officer in the Southeastern Region, had the privilege of working as the I.O. at the 1961 Summer Camp of the Alabama Wing Cadets at Maxwell Air Force Base, and participated in the Search and Rescue Mission for two downed youngsters in the vicinity of Monroeville, Alabama that utilized more than seventy airplanes and thirty three helicopters with the Alabama National Guard and the U.S. Army Aviation Branch assisting.

No, this was definitely a different type of organization. Dave took it upon himself to compose and mail a letter to Civil Air Patrol, National Headquarters inquiring about the renewal of his membership. That is definitely a violation of military courtesy and procedure. The reply stated that there was no record at National Headquarters of his membership, so Dave resigned from the, Senior Pilot Squadron, Civil Air Patrol, Michigan Wing and never again had any association with the organization.

PART XXVIII

A NEW OBJECTIVE

During the first week in August '66 he and Bill Wesley had gone to Rockford, Illinois together to attend the annual Experimental Aircraft Association Convention and Fly-In on Thursday, Friday, and Saturday. While there he saw a restored Great Lakes 2T-1 that belonged to Dr. Dale Drummond of Kansas and examined a "homebuilt" replica of one that was being constructed by the students at the Moody Bible Institute that was on display in a tent. Dave had seen the advertisements in, "Sport Aviation" magazine for copies of the original factory production drawings of the airplane and had paid the advertiser, Harvey Swack three dollars for the, "Info Pack". He also knew that the great aerobatic pilot, Harold Krier flew one that he had modified. Nick D'Apuzzio, the Great Lakes guru in Pennsylvania had designed a biplane based upon the "Lakes" that was flown in world class competition aerobatics by Rodney Jocelyn and Lindsay Parsons that had become known as the PJ-260. The promotional picture in the info pack was of Lindsay Parsons' Great Lakes with an inline six-cylinder Ranger engine on it. It would almost stop any pilot in his tracks. He really was more enthused over the Pitts Special, but Sherry counted the number of seats in both airplanes and said, "If you are going to spend that much money and devote all that time into building an airplane then it must have two seats in it and I only see one seat in the Pitts." During the first week of November Dave mailed a check for one hundred sixty dollars to Harvey Swack for copies of the drawings of the Great Lakes 2T-1A biplane.

The Spirit's Journey

On the day before Thanksgiving 1966 the postman delivered the package of drawings and after dinner Dave opened the package to start examining them. The first thing he noticed was that they had to be organized into the order of the various assemblies such as, rudder, fuselage, upper wing, lower wing, landing gear, etc. After two hours of work he began to ask himself, "What have I done?"

He also thought, "So be it. I've spent one hundred sixty dollars. I can't back out now."

And so Dave began to build an airplane. He often thought, "My co-workers and supervision at Ford probably think I'm insane."

The Federal Aviation Administration and the Experimental Aircraft Association agreed that homebuilt airplanes are built for educational and entertainment reasons. FAA allows it under that premise so long as it is obvious that the builder has performed fifty one percent of the work required for the machine's creation. At the time Dave began construction of his project statistics proved that only one airplane would eventually fly out of every two thousand two hundred projects started. It was an ambitious and daunting undertaking that Dave had started and he knew it.

Dave thought he could do it. He didn't bowl, golf, watch or play baseball, football, hockey, basketball, or seem to do anything that other people choose for entertainment. His wife didn't drink, dance, was very particular about who or what she was sociable with, and was a very adamant and confirmed Baptist. They rarely appeared to have any mutual friends. Their marriage appeared to be one that was unlikely to succeed, but by 1966 Dave was determined to make a home and to maintain its stability for their son, Mike. On April 30, 1967 a daughter, Terri Dawn McKenzie was born in a hospital in Highland Park, Michigan. Terri's birth just set Dave's determination to have a strong and stable marriage "locked in stone".

Dave still remembered his mother's opposition to aviation activities and those people who participated in them, her complaints about the cost, time, and attention that flying demanded and he resolved to not allow flying to destroy his family. He decided that his flying must pay its own way. He'd become a "free lance" Flight Instructor.

On August 28, 1967 he took his first one hour and fifteen minutes of dual instruction in a Cessna 150 from Certified Flight Instructor, Joseph

Frank at Wayne County Flying Service based on Detroit Metropolitan Airport. Joe Frank had been recommended by one of Dave's co-workers at Ford.

It appeared that Joe Frank had scheduling problems and on September 10 he flew with "Dick" Hook who impressed Dave with his very professional manner as a pilot and the debriefing and written handout that he gave Dave after the flight. There was an Instructor. He earned his living as one of the co-owners of Hook's Cleaners in Detroit's western suburbs. On September 16 he flew with a Professional Flight Instructor, Stephen L. Derr who he thought exhibited the standards, manner, and behavior that a flight instructor should always display.

Between August 28, 1967 and April 6, 1968 he had five flights with Joe Frank that totaled five hours of dual instruction and had flown nineteen solo flights that totaled twenty-seven hours and fifty minutes of practice. He didn't believe he was progressing, so he abandoned the school at Detroit Metropolitan Airport and started flying with George Cowles, a professional, full time flight instructor at Mattetal Airport in Plymouth, Michigan. Mr. Cowles gave him five hours and fifty-five minutes of dual instruction and recommended him for the flight test with the FAA. On July 10, 1968 he became a Certified Flight Instructor in Single Engine Land Airplanes with a total flying time logged of four hundred twenty one hours and twenty-five minutes.

At Mettetal Airport everyone congratulated him for having earned the Flight Instructor Rating and when he asked, "Now, do you know of any place where I can find a part time job?" the answers varied from blank expressions to a shrug of the shoulders. The next Saturday morning he drove to the Ann Arbor Airport and received no encouragement for employment in the flight-training field, so he drove south to the next airport which was at Milan, Michigan and introduced himself to the operator, a Mr. Smith. That stop led to a part time arrangement with an extremely small operator on a grass strip with a leased Cessna 150 for a trainer. There were about thirty planes based at the Milan Airport and all of them were privately owned. Dave was told that after he was established he might acquire some work from them. He accepted the opportunity with the agreement that he'd be paid for a flight instructor's market salary of six dollars per hour and the operator would receive the rental for the airplane.

The Spirit's Journey

The next Wednesday evening he was told via telephone to come down to Milan, get the airplane, and fly it to Willow Run Airport, meet a student there and give him a lesson. What a way to start!

Dave drove to Milan, flew the airplane to Willow Run, met the student, flew for an hour, and then flew back to Milan. Over the next seventeen days he flew seven times for a total of nine hours and fifteen minutes in the Cessna 150. Then he started a student from the beginning lesson whom he did later endorse for solo on December 7, 1968 after giving him ten hours and twenty-five minutes of instruction. Ralph Foulke's solo flight was totally uneventful to everyone except him, personally. It was exactly what any flight instructor would hope to see.

By that time he also had a basic student in a Luscombe 8-A, had given a gentleman a check-out in an Aeronca 7-AC, and while on vacation in Alabama given another student one and a half hours of local and two hours and thirty five minutes of cross-country training in a Champion 7-EC. There were three students that owned a Cessna 172 jointly that he was also teaching. His career as a flight instructor was off the ground and growing while he was learning too.

Then on February 16, 1969 while giving a student dual instruction in preparation for his Private Pilot Flight Test in the Cessna 150 that he was using from Milan the engine quit on take off. It was the second forced landing that he experienced in his career. He took the controls from the student and made the landing in a field of winter wheat. After the landing they walked to the home of a neighbor of the wheat field who did not have a telephone but did give them a ride to the nearest airport. From that site they called the operator at Milan and asked him to bring a box of tools.

The operator from Milan disassembled part of the fuel system, found an obstruction, cleared it, and then chose to fly the airplane out of the field. Dave and the student, Ron McCasland rode the van back to the Milan Airport. Upon reaching the airport Dave merely walked into the office, opened the file cabinet, removed his files, and departed without saying anything to the operator, the operator's family, or friends.

Dave then found a flying club at Monroe, Michigan that would allow him to use their airplane and Ron McCasland passed the flight test for issuance of his Private Pilot's license. By that time Dave had four regular students flying Cessna 172's, a nice, simple, basic, four place airplane. Dave looked at the Cessna 172 as a Cessna 150 with two more seats. Of course it really is more than that.

Dave knew he was a flight instructor and now had to control the urge to devote all of his attention to that job. He was already employed in an excellent position at Ford Motor Company, was a home owner, had a wife, a daughter, and a son. He was established in a life and a part time career. Now he had the financial resources to build an airplane that he wanted, but where was the time to build it going to come from?

He'd just have to work on the Great Lakes project when he could and since it was a hobby, just when he wanted to.

In 1969 he had completed the spars for the wings from blanks he had purchased from, Custom Woodcraft owned by Arnold Niemann in Milan, all of the stamped aluminum ribs that he had purchased from Harold Krier, one of the leading air show pilots in the country who had retrieved the dies for the ribs from someone in Cleveland, Ohio where the airplane had originally been produced, and he had made a few of the one hundred ninety two pieces required for the wing fittings from 4130 chrome-molybdenum steel sheet. He had also obtained the rolled aluminum leading edge skins for the wings from the Meyers Aircraft Company in Tecumseh, Michigan, thanks to introductions by an ex-RAF jet fighter pilot, Joe Rayne whom he met at an EAA meeting.

He had redesigned the empennage surfaces to be constructed from cold rolled steel sheet because he wanted a stronger tail than the original design's aluminum tail structure provided and assembled it by tack welding. Then had Oramel Rowe perform the finish welding.

He had also redesigned the upper wing center section to be made from wood instead of using aluminum ribs and trailing edge tube as the factory had. He had also modified the ribs that he purchased from Harold Krier because he wanted to make a minor design change in them to achieve a solution to the rib box cracking problem that the Lakes was known for. Of course it was slow progress, but he didn't abandon his home, family, or career. He didn't wish to make an intolerable job of constructing the airplane. It fit in with the kind of life that he enjoyed and he was enjoying the social life and association with other pilots and airplane builders. The people that he met within the social circle that he moved in were the type people that he tried to become one of. They were conscientious with high moral standards, dependable, productive, reliable, proud, and respectable. He believed he was in the right place at the right time and was happy in it. He hoped he'd fit.

On Sunday afternoon April 26, 1969 he flew Sherry, Mike, and Terri in the Cessna 172, registration number N-5339-R that he purchased a 1/5 share in the ownership of to a privately owned strip located south of Willow-Run Airport named, "Frankman Ranchaero" and made a low pass down runway 27 which put the flight path between the house and the hangar.

The first few times that he'd been to that site Wayne Plattner owned it and was performing the annual inspection on the L-16 owned by the CAP. Mr. Plattner had sold the little field, house, and hangar to Betty and Donald A. Frankman, hence the name change of the airport. As they passed by the house Betty was on the concrete porch at the back door of the house waving and smiling. They landed and were invited into the house and the family felt that they were among friends. Prior to marrying Don Betty Frankman was Betty Skelton, four times United States Feminine Aerobatic Champion and one time World Feminine Aerobatic Champion. They contributed more to the success of one ambitious young man than they'd ever know.

At daybreak on April 29, 1969 he endorsed Dave Nilsson's logbook and Student Pilot Permit allowing him to solo Cessna 172, registration number 5339-R on his sixteenth birthday making it possible for him to claim to be the youngest pilot in the United States for at least a few minutes.

The Cessna 172, N-5339-R was a completely, "instrument equipped" airplane that presented the opportunity for Dave to earn his Instrument Pilot Rating as economically as he could ever hope for, so: he went through the training with Instrument Instructor, Richard S. Johnson and on June 9, 1970 passed the flight test given by Designated Examiner Stewart Peet and added the Instrument Rating to his list of qualifications.

Jack Wells, and Verne Howle passed their Private Pilot flight tests in the spring of '69 and there were several other students whose paths crossed with Dave's, that included Virgil Wolfe flying a Piper PA-12 Cruiser, another "tail wheel" airplane.

Virgil was a mason, a brick layer and he wanted to complete the training for his Private Pilot's License and had been unable to find an instructor to instruct in the tail wheel airplane and also deal with the instrument flying phase of the training using, "partial Panel". Dave "filled the bill" and needed something from Virgil. They made a deal. Dave would help Virgil through acquisition of his Private Pilot Certificate in exchange for Virgil helping him install a foundation, footing, and prepare for pouring the floor of a building that he wanted to erect in his backyard.

The Great Lakes construction project had filled the basement of the house at 26034 Woodbine Dr. in Inkster and Dave was ready to build the airplane factory in which he airplane would be completed. The City of Inkster, Building Department thought it was a garage. The neighbors knew what it was really going to be.

The last flight that Dave made with Virgil Wolfe was on October 21, 1970 in preparation for his flight test where he earned the Private Pilot Certificate. Virgil always seemed to be happy about the outcome of that deal. Dave certainly was. As the years passed and Dave had other experiences and made more observations he decided that Virgil and he never had any idea what a good experience it had been.

Of course there were others who donated time and effort that helped make the, "airplane factory" a reality. Next door was Rodney Beckwith who would often appear with a suggestion or offering of physical assistance like the day that the ten inch steel "I" beams needed to be lifted into place atop the wooden supports Dave had built in the walls. Rod went from house to house in the neighborhood and rounded up guys to gather at Dave's and lift the beams. Gary Tincher was the babysitter's friend who had fallen in love with the street legal dune buggy Dave had built and wanted one for himself. Dave modified the VW "Bug" chassis for Gary's buggy in return for his assistance on the concrete floor. Dave Lucas was one of Dave's students who assisted in installing the aluminum siding on the building. Jerry Walters was the other designer who worked in the same department as Dave at Ford Motor Co. and helped install the dry wall on the ceiling.

In January 1971 Dr. Richard Burlingame from Milan assisted Dave in moving the Great Lakes project from the basement of the house to

The Spirit's Journey

the airplane factory in the back yard and the house appeared to be empty.

Another gentleman who appeared on the scene around Mettetal Airport at this time was Major J. E. "Skip" Kimmerly who was employed by the Michgan Air National Guard at Selfridge Field located northeast of Detroit. Major Kimmerly's job was to test fly the jet fighters that were coming back on line after maintenance. He had built up a "Reed" Clipped Wing Cub, a design that had been a standard in the air-show business in the U.S. since 1939. Major Kimmerly allowed Dave to fly his Clipped Cub solo on April 11, 1971 and later that spring On May 31, 1971 he said, "Dave take the airplane and give your son a ride."

Mike McKenzie was almost seven years old in May 1971 and "Skip" helped Mike into the front seat of the Clipped Cub. Strapped him in, and then "propped" the airplane. Dave took off from Frankman Ranchaero, climbed to about "twenty five hundred feet" altitude and performed a few loops and barrel rolls Mike's mother and sister looked on.

The next day Mike's teacher directed the students to draw a picture showing what they did on the holiday weekend. Mike drew the little red monoplane flying upside down above a field with a hangar and two "stick" figures that were obviously representative of his mother and sister. The teacher found it to be an unbelievable illustration and attempted to assure Mike that it had to be his imagination.

Thanks should be extended to the media for creating and perpetuating the ignorance that the general public supports and maintains about aviation. Teachers read newspapers too.

PART XXIX

BIRDS OF A FEATHER FLOCK TOGETHER

Al Meyers had taken the sketch of the leading edge skins for the wings of the airplane that Dave was building and promised to make a bid for rolling them, so He drove to Tecumseh, Michigan, home of the Meyers Aircraft factory that Saturday morning to receive the results of Mr. Meyers endeavor. The factory was closed as he expected on Saturday morning, so he entered a small door on the side of the building and saw two gentlemen standing at a table and discussing something lying on it.

As he approached they stopped their discussion and focused their attention on each step he took until he stopped in front of them and introduced himself.

"Pard" Diver, the Meyers Aircraft Company plant foreman retrieved the sketch that Dave had given Mr. Meyers and made a few comments and suggestions after explaining that he had been planning to telephone him to discuss Dave's needs and wishes. After agreeing on the design of the parts and concluding their conversation Mr. Diver introduced him to the gentleman who had been standing quietly by and observing the exchange.

He spoke with a very noticeable British accent and promptly explained that the part lying on the table was a part of the Midget Mustang that he was building. He then suggested that Dave follow him to his home and look at his project. There was no requirement that he ask twice.

At his home Dave met Jan who like her husband, Joe Rayne spoke with a noticeable accent revealing her British origin. They had two large dogs and a cat in the house and the two dogs did not seem to care for Dave.

The two men retreated to the garage where the fuselage of the Midget Mustang was standing on its landing gear with the instruments installed, the engine mounted, and the empennage surfaces attached. Dave looked closely at the airplane that was designed in the mid nineteen forties by Dave Long as a "Goodyear Class" midget racer and named, "The Long Midget" which after the designer's death became known as the "Bushby Midget" and later as the "Midget Mustang", model MM-1.

The quality of the workmanship was so good that Dave thought the airplane should not be painted, but should be polished, waxed, and left in its natural aluminum color, had the required identification numbers and placards painted on it, and then flown.

As time passed Jan, Joe, Sherry, and Dave became close friends. Jan especially liked Sherry and Dave's daughter, Terri who had a great love for animals just as she did. Joe soon completed the "Midget Mustang", hauled the airplane on a trailer, ready to fly with a police escort to the Ann Arbor Airport and started flying it.

Joe Rayne was born and grew up in Scotland prior to and during World War II as did Jan. When he became eighteen years old he joined the Royal Air Force where he eventually became one of the pilots of the RAF's first operational jet fighter, the Gloster Meteor. He claimed that his interest in aviation began when he was born and saw the stork fly away.

Before joining the RAF he had built many model airplanes and one design that he especially liked was the "Long Midget" which he modeled, flew, and wished he could have in reality. After leaving the RAF, finding employment in the aviation industry, specifically the Dehaviland Company, marrying Jan, and becoming established in life he began to consider building a "Long Midget".

He soon learned that in Great Britain the government would not allow him to construct anything he desired. The government "approved" amateur aircraft designs for construction by amateurs and "homebuilders" and only approved a few wooden designs that were primarily designed in France. That inspired Joe to move to Canada remaining with the Dehaviland Company where he found limitations considerably less strict, but the Canadian government would not allow him to construct a "Midget Mustang". He acquired employment with the Chrysler Corporation at their test track in Chelsea, Michigan, U.S.A. and established residence in Clinton, Michigan.

He soon found a "Midget Mustang" that was in the process of being constructed and the builder sold the project to Joe. He completed the project knowing that not being a citizen of the United States precluded him from being listed as the owner of an airplane displaying a U.S. registration number.

He kept the airplane at home for three months after its completion until he became a citizen of the United States and then registered it listing himself as the owner/builder and began flying it from Ann Arbor. Sure, there were little problems that revealed themselves during the test program, but he isolated them, identified them, solved them, and the little airplane became a "happy" one.

While he was still in England Joe became aware of a jet fighter designed in Sweden that used the canard arrangement of the wings and control surfaces that was named, "Viggen". He wanted to fly one, but he was British and the airplane was Swedish, so - - -.

In 1971 Burt Rutan presented his design the "Variviggen" to the aviation community and Joe went to London, Ontario, Canada to see the "Variviggen", meet Burt Rutan, and begin serious consideration of owning one. Mr. Rutan appeared with his prototype at the London Air Show that year.

Mr. Rutan had designed the airplane with aluminum wings, wooden fuselage and wing center section, and a short, blunt nose so that it could be built in his garage while he was residing in Kansas employed by Jim Bede. With some advice from Professor Ed Lesher Joe redesigned the entire airplane to be constructed completely of aluminum, built it, and flew it successfully.

Yes, construction of the "Viggenite" as he called it, was a much more ambitious project than the duplication of a proven design. He did have some developmental problems such as, solving a high oil temperature problem, cooling the cylinders on the engine adequately, allowing the engine to receive enough air to produce its rated power output, and solving unforeseen landing gear extension problems. He had built the landing gear and tire wells so small and tight to the landing gear that when it was in the retracted position the combination of tire size increase and landing gear extended length increase over time caused the gear to lock in the retracted position on one flight and he had to land the airplane "gear up". Damage was minor and he soon had it flying again.

During the time that the "Viggenite" was being developed Jan contracted cancer, fought it for a few years, and succumbed to it.

Joe became acquainted with a schoolteacher that he met at the church he attended in Clinton and they eventually married. Jeanne moved into Joe's home located on a privately owned grass runway on the west side of Clinton while Joe continued to test and develop the "Viggenite" which included adding weight in the passenger's seat to investigate the performance envelope.

Jeanne did not like airplanes or aviation, but she had married a pilot. She didn't seem to care to have a runway in the backyard either and unlike Jan, she had no affinity for animals. Jewel would have asked, "Why'd she marry a pilot?"

On the morning of "Labor Day" 1987 Jeanne and Joe boarded the "Viggenite" for Jeanne's first ride in the airplane. There were problems on the take off. The airplane did not seem to have adequate performance and Joe was unable to climb, so he maneuvered around barns, trees, houses, and other obstacles for approximately one and a half miles after lift off. Finally it "dropped out from under him". Jeanne was not injured, but Joe had a fractured back, some vertebrae in the lumbar region, and of course, the "Viggenite" was seriously damaged.

Joe Rayne hasn't flown an airplane since the accident. He and Jeanne have sold the property alongside the runway and built another home in Clinton. Jeanne still teaches school, sings in the church choir, and trades antiques. Joe has served a term as "Township Supervisor" in Clinton, and retired from Chrysler. He has a small machine shop at their home that he uses for his personal "entertainment" and has produced a small jet engine.

PART XXX

N-5747-N AND A LONG TRIP TO FLORIDA

Betty and Don Frankman were executives at one of the largest advertising firms in the Detroit area when Dave met them on April 26, 1969. Betty was a Vice President at tthe Campbell-Ewald Agency in charge of the Chevrolet account and Don, probably best remembered as the Producer of the TV series, Route 66 was a Producer of Commercials for the company. They were a dynamic duo and honestly it was obvious that Betty was the source of ideas, ambition, and energy that generated the forces that drove them. Betty always exhibited a very strong love, devotion, and respect for Don and even though his love for her may not have appeared so obvious he returned it in his own quiet and unassuming manner out of the public's purview.

They had met at the Bonneville Salt Flats in 1965 when Betty was there to drive Art Arfons' jet propelled car designed and built to assault the World's Land Speed Record. She did succeed in the attempt to set a Feminine World's Speed Record. Don was directing and producing a commercial for the Goodyear Tire & Rubber Company. The rest of the story is exactly what might be imagined. They purchased a home in Farmington Hills, Michigan where they resided until '68 when Wayne Plattner decided to sell his home and, airport in Augusta Township, Michigan. They sold the house in Farmington Hills and moved into the two-bedroom home on the grass airport.

Prior to selling the home in Farmington Hills they had purchased the airplane that Betty's father, Dave Skelton had purchased for her in 1948 that she named, "Little Stinker" after the modified Great Lakes 2T-1 that she had named simply, "Stinker" due to its annoying habit of exhibiting numerous mechanical problems and the airplanes shared the unique characteristic of having a swept back upper wing. "Little Stinker" is the second Pitts Special that its designer, Curtis Pitts built and is often mistaken for the first one. The first one was not a very successful airplane and was disassembled. Some of the parts from it are used in "Little Stinker". Ownership of the airplane was the primary justification for them purchasing the small airport.

Early in her air show career Betty had met and become friends with the famous air racing pilot, airplane designer, and builder, Steve Wittman from Oshkosh, Wisconsin. Soon after acquiring the little grass field Betty and Don remodeled the house, rebuilt the 1800-foot long runway, left the 1300-foot long runway alone, and decided that they wanted a two-place airplane in addition to the "Little Stinker".

Betty mentioned that desire to Steve Wittman and being friends since the nineteen forties, Wittman agreed to sell his original, "Tailwind" designated, the Wittman W-8 to them. Betty and Don intended for it to be, Don's airplane.

In August of '69 they and approximately eighteen friends decided to organize a Chapter of the Experimental Aircraft Association and base it at the little airport they had renamed, "Frankman Ranchaero". EAA has a policy of assigning a numerical identifier to each Chapter that is organized and it is done in ascending numerical order. The number for the new Chapter would have been approximately 272, but Betty was the consummate public relations executive and wanted something with some, "pizzazz" in it. The negotiations with EAA led to a compromise. EAA would allow the group to select any number for the identifier higher than the currently available one. The group selected 333 for their identifier and chose the name, "The Flying Stinkers". Don Frankman was elected the first President of the group and Dave was the first Secretary and Newsletter Editor.

In later years the Chapter that had moved to Ann Arbor, Michigan in the '75 to '80 time period changed the name to, "Aviation Pioneers". Dave had decided that he could live with the first name because of its obvious origin, but when the newer group decided to disregard the Chapter's history and heritage he strongly objected.

In January 1971 Dave was elected, Chapter President and soon received a telephone call from the Frankmans. There was a slump in automobile sales at the time and they had decided to leave the industry and move to Winter Haven, Florida. They wanted Dave to know as soon as possible so that there would be time for the Chapter to accommodate the coming changes. Betty and Don planned to depart the last Monday in March and decided to host a party at their home on Sunday afternoon with their friends in Michigan as guests. It was a very good party overriding the feelings of loss overhanging the gaiety.

Guests included executives from General Motors such as Zora Duntov, father of the Corvette, Bill Barber, airline Captain and world famous aerobatic pilot, Fred Letau, Don Ziegler, builders and pilots of experimental airplanes, and Dave McKenzie, flight instructor and designer from Ford Motor Co. There may have been fifty guests sitting on the carpeted floor in the empty living room nibbling hor-d'ouvres and drinking champagne from paper cups.

Thursday afternoon June 17, 1971 the telephone rang in Dave's office and upon answering he heard Betty say, "Barber's leaving Detroit tomorrow morning to fly, "Stinker" down here for us. Would you like to bring, "Sylvester"?" Steve Wittman's first name really isn't, "Steve" and he didn't appreciate Betty painting his name, Sylvester on the "Tailwind" accented with a giant sized painting of, Sylvester the cat on the vertical fin.

Due to both airplanes being at, Frankman Ranchaero with its short, grass runways and, "Stinker" having no room for Bill Barber's baggage he and Dave agreed that he would fly, "Sylvester" to Mettetal Airport that afternoon and Dave would pick up Bill's baggage when he drove him home. The check out in the airplane by Bill consisted of showing Dave how the fuel system worked and then saying, "She's a little pussy cat. Go fly it some and I'll wait at the picnic table over there."

Dave flew the little airplane out to the west of Mettetal Airport and performed a series of stalls and some slow flight. Then he returned to Mettetal Airport where he made two landings and during the drive to Bill's home they agreed that Bill would call him on the phone at Mettetal when he was ready to take off in, "Stinker" the next morning

The Spirit's Journey

and Dave would meet him in the air over, Frankman Ranchaero. He would take off when he saw Dave in the air.

Upon arrival he could see the Pitts sitting on the ground at the house and as he circled Bill came out of the house followed by Shirley Pargo, the new owner's wife. Dave circled overhead while Shirley stood beside the plane and Bill propped it. They obviously had some difficulty starting it, but it finally started and then stopped. The next time he flew past Bill gave him a wave off, so he went to Willow Run, landed and then called the Pargo home on the telephone.

Shirley answered and explained that Barber had shown her how to handle the switch and throttle while he hand propped the airplane. She had never been around an airplane, but did as directed while standing beside, "Stinker". When it started she made a sound as though startled. The dog that was standing nearby apparently thought that Barber had threatened her, so it responded by biting Barber on the leg. Barber decided that a visit to a physician was in order and told Shirley to advise Dave that he'd call when he was ready to depart.

The initial take off at Mettetal had started about 8:30 that morning and it must have been 2:00 in the afternoon when Bill's medical needs were met, completed, and Dave met Bill above Frankman Ranchaero with Bill flying the lead to Chillicotti, Ohio.

They were soon separated because during the pre-flight inspection Dave had decided not to fly the engine at the 2700 R.P.M. recommended by Steve Wittman. The four-cylinder engine obviously could tolerate some maintenance and Dave was not comfortable with the idea of abusing it. It obviously seemed to run its smoothest and quietest at 2425 R.P.M. and at that power setting the, "Tailwind" was cruising at 132 M.P.H. Besides, Dave enjoyed flying. He was in no hurry.

The Pitts that was smaller and lighter than the later model Pitts's and instead of having its original 90 H.P. engine had a 160 H.P. Lycoming installed on it by one of its prior owners. It must have been cruising 145 to 150 M.P. H..

The airport they had chosen for a fuel stop at Chillicotti was closed, but not marked as such at the site or on the chart, so Dave landed and found it closed. He then took off and backtracked to the preceding airport and found Bill and the Pitts on the ramp at the fuel pump with Barber already on a first name basis with everyone at the airport. The logbook entry indicates that they had flown for one hour and forty-five minutes to travel from Willow Run to Chillicotti.

While the airplanes were being refueled they decided that the next stop would be at Wise, Virginia. Because neither of them knew how much fuel the Pitts carried or what its consumption rate was they did not dare expect it to remain airborne more than two hours. Dave was to fly the lead on that leg and Bill agreed to make certain to remain in formation with him.

Dave took off with Bill following and there was nothing particularly memorable about that leg of the trip except when passing a towered airport south of Chillicotti Dave couldn't get a call in to the tower as early as he wished because the controller was using the tower, ground, and clearance delivery frequencies at the same time to deliver an instrument clearance to an airliner that was on the ground. Dave did bend to the west. As they were passing the airport at Whitesburg, Kentucky he was growing concerned about the fuel supply in the Pitts. They landed at Whitesburg and checked the fuel in the Pitts, decided it was O.K., couldn't see anyone at the airport, and continued to Wise, Virginia.

The weather was marginal VFR because of visibility, so the Pitts took off first and promptly disappeared in the haze. After Dave took off he circled the airport to be sure that the Pitts hadn't suffered fuel exhaustion and gone down. Not seeing any sign of the Pitts he continued on to Wise and was relieved to see the Pitts on the parking ramp. The flying time from Chillicotti to Wise is logged as two hours. They were then given a ride into town by a couple that just wanted to do a favor for the pilots in the strange airplanes.

They spent the evening in an ancient hotel that had originally been named, "The Inn at Wise Courthouse". They could have stayed in a row of cement block motel rooms in the back yard, but both of them thought they might never have the opportunity to stay in a 1770 vintage hotel again. Dave found dinner and a short walk afterwards to be an absolutely memorable one with Bill Barber telling stories the entire time. Bill was an unsurpassed raconteur and that was one time he had an audience of one hanging onto his every word. There were air show stories, airline stories, and several times he would take a break between stories to speculate on various means of acquiring vengeance on a dog that would bite a man that was propping an airplane.

The weather was still marginal the next morning. After breakfast and in the Inn at Wise Courthouse they decided to fly through the Appalachian Mountains by following a connecting group of valleys that also had a railroad track, the "iron compass" on their floors. There were numerous sidetracks branching off and running up valleys probably to mines, but they thought those should present no problems. After planning the route to Greenwood, South Carolina they asked one of the local pilots at the airport to review it to take advantage of his expertise and knowledge of the area. Bill took off first and performed some Cuban eights, slow rolls, and loops as thanks to the staff at Wise while Dave climbed out in the Tailwind.

They left Wise with Bill flying "lead" and were having no problems. They were in a tunnel formed by mountains on each side extending into the cloud deck above them when Bill made a left turn following a railroad track. Dave thought, "WRONG TRACK, BILL", but followed him and very soon both of them knew they were not following the correct track. Bill started a steep banked turn to the right and Dave did a wingover to reverse course. In a few moments they were back to the correct track and Dave, knowing that Bill would turn left turned right to let him pass. Then Dave performed another wingover and "Stinker" had disappeared in the haze.

Yes, the visibility was poor with a low ceiling and mountains all around. He'd been there, done that before. "Stinker" was being flown by a professional airline pilot, air show pilot, Captain of the United States World Aerobatic Team in 1962, and a natural pilot if there is such a being and they had plenty of fuel, so Dave wasn't concerned. He flew on to Greenwood, South Carolina and could see the Pitts parked on the ramp before he entered the traffic pattern.

When Dave walked into the office he found Bill listening to the owner of the operation recounting how he had once performed an oil change on, "Stinker" when a little girl had landed there. Obviously the consummate raconteur and communicator was doing well, but this time he was listening. It had taken them two hours to fly from Wise, Virginia to Greenwood, South Carolina. Flight planning had been just fine and the flying time from Wise to Greenwood had been exactly the planned two hours.

Due to the little difficulties and inconveniences encountered thus far in the trip Dave hadn't been really enjoying it, but he led the flight to Waycross,

Georgia and another fuel stop. He then led another flight to Leesburg, Florida where they stopped again for fuel and the final leg to Winter Haven, Florida. The total flying time from Greenwood, South Carolina to Winter Haven, Florida was four hours and twenty-five minutes.

There were times on those legs of the trip when he'd wonder, "Where's Barber now? I haven't seen him for several minutes." He'd raise the left wing and yaw the airplane to the left while attempting to look behind and see Bill go by his left side inverted and Bill waving a hand, then half rolling to upright and dropping into position behind him. Sometimes Bill would go by on his left and below, then pull to the vertical and roll going up. He suspected that he would have a mental picture of those scenes in his mind for the rest of his life. . .

What a totally unique experience! Dave realized that he had been seated in a grandstand seat that was a museum piece of aviation history observing one of aviation's big names performing in another museum piece of aviation history. There have not been nor are there likely to ever be many more pilots who ever had or will have such an opportunity. Dave felt humbled and simultaneously honored to have been allowed to fly the little red monoplane that he saw in Steve Wittman's hangar at Oshkosh, Wisconsin sixteen years earlier across the United States from north to south

"Sylvester" and "Stinker" on the ramp for fueling at Greenwood, South Carolina (author's collection)

The Spirit's Journey

Steve Wittman's given name was, Sylvester. (author's collection)

Sunday evening June 20, 1971 Dave was sitting in the living room of the Frankman's home in Winter Haven with about ten people scattered around. Betty and Don were hosting another party. Suddenly Betty jumped up and said, "Don, call Jack Brown. Dave, give me that drink. You're flying tomorrow. You're the only pilot in the room without a Seaplane Rating."

The Frankmans were going to drive Bill and Dave to Tampa the next morning to ride an airline back to Detroit, but the group had been discussing seaplane operations and experiences when Betty noticed Dave's silence. He tried to be quiet and to listen on most occasions, but would let his guard down and open up when the subject of aviation was broached. Perhaps that's the reason that Betty, the fabulous hostess noticed him and it occurred to her that his lack of seaplane experience was the reason for the silence.

Don came back into the room and announced that He would drive Dave to the "Jack Brown Seaplane Base" Monday morning and then drive Bill to Tampa. He and Betty would drive Dave to Tampa when he had the Seaplane Rating in his pocket whenever that might be.

The first two hours of instruction was from Jack Brown, the owner of the school and the FAA Designated Examiner who gave flight tests and if the applicant passed issued the rating. Gordon W. Currier, a retired U.S. Navy pilot who as a student was the second person to land an airplane at the site that became the U.S. Navy, Great Lakes Naval Air Station near Chicago, Illinois, gave the next two hours and twenty minutes of instruction. His certificate number was, CFI 9307. It's probably the lowest license number in Dave's collection of endorsements.

Late that afternoon Jack Brown rode with him for the flight test and issued the Rating. Now Dave held Commercial Pilot with Instrument Pilot and Seaplane Ratings on it. He also held the Certified Flight Instructor certificate with authorization to teach in land and seaplanes.

That night Betty and Don drove him to Tampa, treated him to dinner at the Columbian Restaurant in Ybor City and bid farewell at the airport. He arrived at Detroit Metropolitan airport around 1:00 A.M. Tuesday morning and reported to work in the, "Vehicle Assembly Simplification & Serviceability Research Department" in the Scientific Research Laboratories of Ford Motor Company at 8:00 A.M.

PART XXXI

"CLIPPED WING" CUB, AGAIN

In Experimental Aircraft Association, Chapter 113 there was a couple, Joann and George Ardwin that other members made no attempt to understand or explain why they loved each other and married. George earned a living driving a truck delivering milk to stores and restaurants in the early morning hours. He was a "milkman" who liked to fly.

Joann was an absolutely stunningly beautiful young lady that had been a waitress in a donut shop that George frequented on his milk route. She also understood how to dress in the latest and most appealing manner regardless of the time of day, night, or occasion. The ladies associated with the Chapter included at least one who was a member and of course, several wives that never expressed any opinions or thoughts about the union of the Ardwins, but it is certain that Dave wasn't the only guy in the organization that wondered how on earth George ever attracted a lady of Joann's beauty, poise, taste, and grace. Sherry, Dave's wife would attend the EAA 113 meetings the third Thursday evening of each month and by this time Dave was instructing so much that he'd forget about having part ownership in N-5339-R, the Cessna 172-F in the hangar at Mettetal Airport.

During some of the meetings the thought would enter his mind that it would be nice if they went flying after the meeting. He'd mention it to Sherry and she'd ask the Ardwins to accompany them. They became friends, were invited to the Ardwin's home for dinner, and Dave decided that Joann was a talented cook too. George was lucky. He certainly didn't seem to have any sex appeal. Of course, no other homo-sapien of the male gender ever appeared to have any sex appeal to Dave either.

One evening George announced to Dave that he'd bought "Skip" Kimmerly's Clipped Wing Cub and wanted a "check out in it before attempting to fly it from, Frankman Ranchaero to Big Beaver which was the airport closest to his home.

On July 11, 1971 Dave flew with George in N-26-SK and in just twenty minutes decided that George could safely fly the airplane. George left Frankman Ranchaero and flew the Clipped Wing Cub home to Big Beaver where he tied it down outside on the east side of the airport with its nose facing west and the tail pointing towards the homes that had been built with their backyards next to the airport property. Within a month high winds accompanying a thunderstorm tore the airplane loose from its tie downs, lifted it high in the air, turned it onto its back, and dropped it through the roof and into the dining room of the nearest home.

The story really tugged on Dave's heartstrings. He never believed the little airplane should have been left outside, but George owned the airplane. It was his, not Dave's, so it really wasn't any of his business.

Dave didn't see Joann and George again after that disaster and later heard that they had divorced. He thought they were both much greater losses than any man should have to endure in one lifetime. He was to learn more about that at a later date.

PART XXXII

A NEW OBJECTIVE (CONT'D)

Perhaps it was Dave's imagination or maybe he suffered some paranoia, but it seemed to him that anytime an acquaintance or co-worker asked for a progress report on the "home-made airplane project" he'd hesitate to answer. He knew he couldn't produce obvious results on the Great Lakes without working on it and he couldn't work on it without having the tools and materials and he couldn't have those without having the money that he couldn't have without the flight instructing and he couldn't do that and work on the airplane at the same time. Reality is sometimes the creator of a brutally vicious circle. Sometimes Betty Frankman would ask why he didn't work on the airplane and point out the obvious truth that he'd never finish it if he didn't do something on it. She was a, "pusher" and could produce verbal incentive. He didn't resent it though. She was right.

One of her questions was, "When are you going to start on the fuselage?"

In 1970 he had made contact with a company in New Jersey that advertised precut and trimmed tubing packages for various homebuilt airplane fuselages and their response had indicated that they had never produced a package for the Great Lakes, but they were currently holding orders for two sets and if Dave placed his order they would provide the three packages in fourteen weeks from the date of the order. Dave mailed a check.

A year later he was becoming disillusioned about the company and its promises and his letters of inquiry for status reports on the order were beginning to show his displeasure.

One and a half years after placing the order with Kerbie –J, he received the package that was completely and totally satisfactory.

He then built wooden frames at each of the fuselage stations and assembled the tubing package inverted on the tabletops that were the basis of the assembly fixture. (No, it was not an error.) Then he tack welded the package together and when the family was going to make its annual trip to Alabama to," renew the animosity with the relatives" he tied the assembly on the top of the '70 Mustang fastback coupe.

At the, Ed E. Reid Vocational College in Evergreen, Alabama where he had assumed the responsibilities of, Director of the Welding Department and taught the skill to students enrolled in the course, Mac spent about three hours per day for the next two weeks welding it together by the tried and true oxy-acetylene method. At the end of the two weeks Mac and Dave tied the fuselage back on top of the Mustang. There were many odd expressions on the faces of passengers in passing automobiles during the return trip to Inkster, Michigan.

Great Lakes upper wings under construction in David's basement. (author's collection)

The Spirit's Journey

Mac performing the final welding on the Great Lakes fuselage. (author's collection)

Great Lakes fuselage ready for return to Michigan from Alabama (author's collection)

With the fuselage finally on the gear, that is, standing on its own wheels it was time to mount the engine. Jennings Carter, the "duster pilot" in Monroeville, Alabama had donated a run-out Continental W-670, seven-cylinder radial engine for Dave to use as a building fixture because that was the engine that he had selected to power the, "Lakes".

The source for the drawings of the airplane was still active and he was selling drawings of various modifications that he claimed were proven modifications, so Dave purchased the drawing of the W-670 Continental engine mount. Dave immediately noticed features about the drawing that raised serious questions. He talked to the source of the drawings and was referred to the Eastern Airlines Pilot who had made the drawings.

Bob Rust said, "That engine installation was done on my Fleet biplane. Harvey saw it and said, ""That's about the same size as the Great Lakes and should work on it too."" Dave performed his own installation design featuring a swing-out engine mount and never had any doubts about it working fine. The installation of the 220 H.P. radial engine instead of the original factory produced 90 H.P. Cirrus, four-cylinder, inline engine naturally led to a complete redesign of the fuselage turtle deck, stringers, side panels, and fairings.

Perhaps the empty basement no longer appeared, normal without being full of airplane parts, so he purchased a, Woody Pusher project from Don Smith because it was similar in appearance to the Curtis Junior pusher that his father owned in 1935.

Dave's own flight training hadn't ceased. He continued to pursue additional experience and additional ratings. The next obvious rating to add to his resume' was the Instrument Instructor rating and on September 14, 1973 he took the first lesson from Don Duff in a Cessna 172 based at Mettetal Airport. On some occasions the airplane failed the pre-flight inspection by Don and Dave. Don would be greatly annoyed with that development and would suggest that they use, "an airplane that was maintained". They'd fly his Beechcraft Bonanza on those flights with Don sitting in the left seat and Dave trying to instruct from the right seat. Don had the dual control yokes in his airplane, but did not have brakes in the right seat position. That was O.K. because Don in the left seat with a full set of controls was the, "pilot in command" on take-offs and landings.

On November 5, 1973 Dave passed the oral examination and flight test for the Instrument Instructor rating given by Inspector Leland H. Gunther of the Detroit Flight Standards District Office of the FAA flying N-5339-R, the 172 that he owned 1/5th share in.

During one of the annual visits to south Alabama he excused himself from the task of renewing the relative's animosity for five days and visited Panama City, Florida, the home of Sowell Aviation and on December 29, 1974 added the Multi-Engine rating to the list on his Commercial Pilot Certificate. William Sowell, Jr., the son of the man who gave him his Private License over fourteen years before conducted the examinations.

PART XXXIII

A FEW TALENTED YOUNGSTERS

June 15, 1971 is the date of the first appearance of any note mentioning a Champion 7KCAB, registered as N-6389-N in his logbook. That was the date of a "test flight" with Dr. Alexander J. Kovach who was looking for a flight instructor for his 15 year old son, Kim Kovach. "Doc" accepted Dave as a suitable instructor and a long relationship was in its budding stages. "Doc" and a friend had purchased 6389-N from an air show pilot named, "Cowboy Bob" Carter and then it sat in a "T" hangar at Mettetal Airport while "Doc" and his partner attempted to resolve some differences. Those differences were resolved when "Doc" purchased his partner's share in the airplane and on July 7, 1971 Kim received his first hour of dual instruction from Dave.

Kim was another challenge for Dave. He was another "natural" pilot and Dave was faced with a challenge to keep the training syllabus interesting to Kim while waiting for his 16th birthday so that he could solo the airplane.

His first solo occurred thirty minutes after sunrise on the morning of September 25, 1971. It was uneventful except for the local police arriving to see what was going on when they saw two cars and several people milling around an airplane and hangar before sunrise.

There was a column named, "Flight Lines" published in the Detroit News every Thursday evening and Kim was surprised to find that his achievement was mentioned in the Detroit News column called, "Flight Lines" by Al Lowman the next Thursday. He was the youngest pilot in the United States, at least for a little while when he soloed. Kim passed his oral examination and flight test for the awarding of his Private Pilot's Certificate on November 15, 1971 and like Virgil Wolfe, had to perform the instrument flying part of the test on partial panel. That test flight was executed after Dave flew with the examiner to check him out in the airplane prior to him testing Kim in it.

Dave enjoyed the privilege of training Kim for his Instrument Rating and Commercial Pilot's Certificate. Additional training that Kim, received from Dave was in aerobatics in his fathers Citabria that Kim had recovered and repainted by that time. It was without a doubt and inarguably the flashiest Citabria ever built when Kim completed it.

Kim became a Flight Instructor, graduated from college as an accountant and is now the Controller for an auto parts supplier in Plymouth, Michigan.

John O. Maxfield's father had been taking him to the airport regularly since he was four years old. His father, Orlo owned a Funk that has the distinction of being the second to the last airplane that the Funk brothers produced in Coffeeville, Kansas. It was not the first Funk that Orlo had owned. The Funk airplanes had been designed and production begun in the mid nineteen thirties in Akron, Ohio and were powered by a modified Ford, Model B automobile engine. Orlo had owned one prior to entering his country's military service at the beginning of World War II. and was the second owner of N-1654-N. The Funk brothers were the first owners. It was the next to last Funk constructed.

John had his first flying lesson with Dave from Grosse Ile Airport on December 3, 1972. It was immediately obvious to Dave that this was not the first lesson that John had received and neither John nor his father ever disclosed who had been giving him dual instruction, how much of it, or any other information about it, but it's difficult to conceal knowledge and ability from someone who also knows the subject matter.

John's first lesson from Dave was received in a Cessna 150 December 3, 1972 and he continued to use the Cessna 150 until his father purchased a half interest in a Cessna 170 that John continued his training in and on August 7, 1973 transitioned into the Funk that he made his first solo flight in on August 7, 1973. It was another first solo as young as the pilot could possibly legally do it. This time the airplane was almost twice as old as the pilot and Al Lowman at the Detroit News announced that one in his column too.

Dave had attempted to convince Orlo that his son should wait until he was 15 ¾ years old before beginning the training, but Orlo would not accept that recommendation. He wanted John to start, "NOW". It was another challenge to the instructor. How do you keep it interesting to the student? It's difficult to stay on a logical syllabus and not have the student get bored or become discouraged, but it obviously worked out fine.

August 1974 the annual EAA Convention at Oshkosh was going to open and the AC Spark Plug Division of General Motors sponsored a rally with starting points scattered around the United States for "Oshkosh" attendees to participate in. The rally required that each participant competing in the event provide their estimate of the fuel consumption for the flight and the speed that they thought the airplane would make on the trip. Orlo and John approached Dave about the possibility of a Student Pilot, especially John Maxfield flying an antique airplane in the AC Spark Plug Flight Rally to Oshkosh, Wisconsin from Flint, Michigan.

A student pilot was and is still required to have the endorsement of his/her flight instructor in their logbook approving the flight prior to attempting it. Dave agreed to provide the endorsement provided John flew the airplane from Grosse Ile to Flint, REFUELED the airplane, and then started in the rally after obtaining a weather briefing from an FAA Flight Service Station the morning of the departure. Dave knew that Orlo would be flying the Cessna 170 on the same route at the same time. There did not appear to be much opportunity for errors to be made.

The flight from Grosse Ile to Flint in the Funk went well, but John didn't refuel the Funk upon arrival that afternoon. He decided to refuel the next morning to avoid temperature differences between the two days causing fuel consumption measurements to vary. Orlo said, "Oh John, just go ahead. You're not going to win anyhow and that Funk has plenty of gas."

The three most useless things in aviation are 1. Altitude that's above you. 2. Runway that's behind you. 3. Fuel that's on the ground.

The Spirit's Journey

John allowed Orlo to talk him out of refueling and start the flight without a full tank of fuel. He landed in LaPorte, Indiana and checked in with his time and there was a line of contestant's airplanes at the fuel pumps. While he was waiting in the line for fuel Orlo arrived and said, "Oh John, you don't have a chance of winning anyhow and that Funk has plenty of fuel to make it to Morris, Illinois. Go ahead." The sixteen year old Student Pilot, John followed his father's instructions and departed with the destination of Morris, Illinois planned for his next stop.

Approximately ten miles before reaching Morris, Illinois the fuel was exhausted and the engine quit running. The Student Pilot, John made the forced landing in the antique Funk airplane on U.S. Interstate Highway 80.

While the Illinois State Police were attempting to decide how to handle this situation an airline pilot that they were acquainted with arrived. The airline pilot owned a private runway at his home located close by, so he went home and brought back five gallons of aviation fuel. He and John poured it in the airplane and the State Policemen agreed to allow John to take off on the expressway if the airline pilot would go with him. Problem solved. The airline pilot advised John that he had never seen that type airplane before and would be more comfortable if John performed the take off, the flight to Morris, and the landing.

After the refueling at Morris it was obvious that John had no chance of completing the rally in a competitive position, so he bypassed the planned stop at Waukesha, Wisconsin and flew on to Oshkosh where Dave met him.

Dave had driven his Mustang to Oshkosh with Sherry, Mike, and Terri and towing a camper trailer to spend the week in. Kevin Kovach, Kim's younger brother met him at the campsite and advised that he had heard that John Maxfield had some difficulty on the flight from Flint. Dave investigated the story, learned of the fuel starvation problem and went to the FAA's tent at the convention site.

He discussed the events as he thought he had heard them and made arrangements with the FAA Operations Inspector for John to visit him the next morning. He and John visited the Inspector the next morning and of course, everything was O.K. It had been a good training exercise, but not one that he recommended being practiced regularly.

On the return flight to Grosse Ile John stopped at the FAA Office on DuPage Airport and asked for the Inspector that he had met at Oshkosh and discussed his fuel consumption problem with to sign his logbook indicating that he had successfully made one of the planned stops enroute home.

On his seventeenth birthday, which is the earliest that a person may be issued a Private Pilot Certificate, August 24, 1974, John passed the Oral Exam and the Flight Test for the Private Pilot Certificate flying the Funk. The instrument flying portion was performed using, "partial panel".

Dave enjoyed the privilege of training John for the instrument rating in a Cessna 172 based at Detroit Metropolitan Airport in Detroit, Michigan. John went on to earn the Airframe and Powerplant Mechanic Certificates, the Multi-Engine rating on his Commercial Certificate and then the Airline Transport Pilot Certificate with several airplane type ratings on it. He became a professional pilot flying for Chrysler Corporation until the early eighties when the company had to divest itself of the Aviation Department.

The department was renamed, "Pentastar Aviation" and John remained with them approximately twenty-five years and at the time of severing his employment he was a Captain on their Rockwell American "Gulfstream" G-5 that he regularly flew from Pontiac, Michigan to Frankfort, Germany and return. There was one flight that he made that took off in Pontiac, landed in Frankfort, then flew to Tokyo, Japan, and then returned to Pontiac, Michigan.

Imagine an ordinary business tool that is capable of circumnavigating the planet with only two stops.

Another student during that period was Karen Jones whose father, Larry Jones owned a Cessna 150 that he based at Grosse Ile. He was hoping his daughter would solo on her sixteenth birthday, but the weather precluded that possibility on her birthday and she had to solo on the day afterwards.

Karen never completed the training for the Private Certificate, but she did graduate from college with a degree in Electrical Engineering, is married, has a child, and is an engineer at Ford Motor Company.

There was also Ruth Gilmore, a sixteen-year old young lady whose father was a school teacher who owned a Cessna 150 and a Cessna 172 wanted Dave to train her. The training began on May 27, 1976 in the Cessna 150 and on

The Spirit's Journey

July 21 she made her first solo flight and he signed the recommendation for her flight test for the Private License on March 11, 1977.

Her father, Howard Gilmore had been struck by a passing car while he was trying to enter his car after work at the Stout High School in Dearborn, Michigan and was unable to appear for work as the Vocational Trades Instructor, so the day after Ruth passed the flight test they left Michigan to go to California in Howard's Cessna 172 with Ruth piloting the plane. (Howard was navigating and assisting with other responsibilities.)

Early in 1973 John Maxfield had begun to visit Dave's home at least two nights during the week and on most Saturday evenings assisting Dave on the construction of the Great Lakes. He never asked for it, but Dave quit charging for the instruction he was giving John. Ruth's home was in Garden City and she started making regular visits in '76 when she would bring her flight planning charts, computer, plotter, and forms over to plan her solo cross country flights. Of course she'd get involved in something relative to the Great Lakes on each visit.

John jokingly referred to himself, as the, "grunt" working on the Great Lakes and when Ruth became a regular visitor he'd say, "Here comes the, "beaver grunt." She knew of it and neither of them ever made any issue of it.

John decided that he wanted a Pitts biplane and Dave knew that Bob Lyjak, the air show pilot and mathematics Professor at the University of Michigan, had a fuselage truss for one that had been fabricated by Oramel Rowe, the expert welder and airplane builder who resided in Rawsonville. John and Dave visited Bob, inspected the fuselage, and with help from his father, Orlo John purchased it. They took it to Dave's home and it joined the Woody Pusher fuselage in the basement. The basement then presented a "normal" appearance.

John then purchased the wings from John Gardner's Pitts when Gardner replaced them with a set of "symmetrical" airfoil wings that he purchased from Henry Haigh. The wings were then hung in the ceiling above the fuselage. Dave visited Henry Haigh, measured the aluminum landing gear on his "Super Pitts" and made a drawing of the landing gear. Orlo and Dave combined resources and had the landing gear made by a company in California to present to John for a Christmas gift and John's Pitts project was beginning to look like a serious project.

PART XXXIV

"THE SIGN", ATC, & A TRIP FROM NASHVILLE

What is happening on the planet, Earth? What became of integrity? Has selfishness, indifference, and greed become the dominant trait of society? Does the concept of retreat into isolation have some appeal? Perhaps those questions, when answered may lead to comprehension of the reasons that Dave enjoyed solo night flight, even single engine.

When Bill called his interest was instantly aroused. "Dave, you wanna' bring my plane home from Nashville?" was the question he heard when answering the telephone.

Bill Unger and Dave had been friends since '71 when they met at the Frankman's home. Bill and his son, Danny who already had a Private License and was preparing to leave for college had driven out to the little airport to see what was going on and Bill learned that Dave was a flight instructor.

Bill had joined the MANG Flying Club based at Detroit Metropolitan Airport and asked Dave to teach him to fly in one of the club's planes. "MANG", was the abbreviation for, Michigan Air National Guard. It started a long and close friendship. Dave had met Bill's wife who was from a small town near Meridian, Mississippi and Dave felt that because of his parent's heritage he and Bonnie shared a common link. Dave liked Bonnie in spite of the age difference. She wasn't old enough to be his mother and was too old to be his girlfriend.

Bill had earned his Private License and had owned two other airplanes prior to purchasing the Cessna 150 that he wanted Dave to retrieve from Nashville, Tennessee.

Dave explained that Sherry was in Georgia renewing her relative's animosities and their children, Terri and Mike were at home with him. Bill solved the problem by volunteering Bonnie's services to take care of them for one day, so Dave arranged to take one day of vacation from the, High Temperature Gas Turbine Engine Department at Ford, his latest assignment in the ongoing series of reorganizations in the company.

With all the obligations and logistics met Bill purchased a ticket on Southern Airways from Detroit to Nashville, picked Dave and the children up early the next morning and left Dave at the Terminal on "Metro" with a promise to pick him up after receiving a call advising that he was at home with the 150.

Dave boarded the DC-9, later called an MD-80. He decided to attempt to wrangle an invitation to visit the, front office, so on the back of a photo of the Great Lakes project he wrote, "I'll let you fly mine if you let me see yours." gave the photo to the stewardess and asked her to give it to the Captain. Airline security was becoming a huge concern due to the "D. B. Cooper" hijacking in the northwest and several others by unsuccessful people in the United States wishing to move to the island of Cuba. She was soon back at his seat and said, "Sorry, as much as he'd like to he can't. Security won't allow it."

Of course he was served a breakfast and arrived in Nashville about 8:30 in the morning, was met by the person in possession of Bill's 150 who announced that he and two of his employees needed to stop for breakfast before going to the shop.

During breakfast Dave learned that the airplane was in Nashville to have a sign attached and installed. The proposed sign would have looked like a part of a wire fence laid horizontally and attached with legs to and beneath the airplane. The right seat would be removed and was to be replaced with a metal box that had an open tray in the bottom with a roller extending into the tray and another roller driven by an electric motor at the top. The "fence" was made of strands of wire spaced at 10 inch intervals running span wise and additional strands running fore to aft spaced at 10 inch intervals. The overall dimensions of, the "fence" were 50 feet wide by 10 feet long. At each intersection of all the wires there would be a small reflector and an electric light bulb making the sign visible at night from below the airplane. Of course there was a separate wire running to each light bulb and all of the wires made up a bundle about eight inches in diameter that entered the fuselage through a hole cut in it behind the right door and connected to the metal box mounted on the right seat tracks.

The sign functioned by punching holes through a fabric reinforced paper tape. The programmer would punch a pattern of holes through the tape that would simulate letters of the alphabet spelling out the message to be displayed on the sign. The operator would then thread the tape around the rollers, fill the tray in the bottom of the metal box with liquid mercury, and the sign system would be loaded. The bundle of wires entering the fuselage through the baggage area would carry the electrical current that was completed by feelers on their ends that would make contact with the liquid mercury in the tray at the bottom of the metal box, thereby completing the circuit.

Then when flying above a gathering of people at night the pilot could turn on the electric motor in the "metal box", turn on the sign lights, and the device would display a moving, ever changing, sign similar to the one seen in movies on the building in New York City and on the side of the Goodyear Blimp.

After breakfast they were riding north bound on a road on the east side of Nashville International Airport when the boss said, "There it is". Dave spotted the airplane parked under a tree beside a dilapidated house on the opposite side of the road from the airport. It was obvious that they had taxied the airplane off the airport, down the public street, into the gravel driveway, and parked it under the big tree where the modifications were being made. There was no sign on it and it was already 10:00 A.M.. He thought, "Wow! Now this is really a, shade tree operation."

The gentleman who was driving the car said, "She's not quite ready yet, but we'll finish her today." The seats were not in the airplane. The doors were off. The radio was missing and there was no indication of it being ready for the installation of a sign. Bill's son, Danny had flown the airplane to Nashville from Detroit and prior to his arrival he had found that the drive belt in the radio connecting the tuning knob to the frequency read out drum had broken. The radio operated O.K., but it wasn't possible to determine what frequency the VOR receiver was tuned to. There was no reason to expect navigation with that radio to be anything other than difficult, at best.

Late that afternoon they taxied the airplane, up the street and onto the airport. The FAA Inspector checked over the installation of the sign and "signed it off", which is aviator language for, "approved it". He also issued a RESTRICTED category license on the airplane. The restriction was for the pilot to be the only occupant of the airplane when it was flown. Now, Dave thought, "I can go home."

The gentleman that seemed to own and supervise the business that had installed the sign said, "Now, we'll check you out in the airplane."

Dave asked, "How?"

"You sit in the baggage area and I'll fly it around one time. That way you can see how we do it." he casually ordered.

Dave was simply astounded. Number one, a RESTRICTED category airplane, "PILOT ONLY" being the restriction means exactly, "No one aboard in flight except the pilot." Number two Dave was to ride in the baggage area with no seat belt. Number three, he was to observe the pilot and by so doing, the pilot is going to decide whether or not Dave is competent to fly the airplane back to Detroit.

While using his restraint, Dave replied, "If you want to know whether or not I can fly it, then stand here and watch me take off. For observation of the landing, I'll tell you where my first fuel stop will be and you can go there and watch the landing." He never saw the man again.

He ordered fuel for the airplane, loaded, and took off.

With darkness approaching He felt that he really should attempt to resolve the radio problem. He simply turned the frequency selector knobs at a ratio of, the inner knob ten clicks to one click on the outer knob until the course deviation indicator responded to a signal. Then the question became, "What station is this, hence what frequency do I have?" He couldn't read Morse code by ear, therefore identifying the station had to be done the hard way. He knew his location, so he turned the omni bearing selector until the CDI (course deviation indicator) centered, then looked on the chart (map) in the indicated direction until he found a VOR station indicated. London, Kentucky seemed to be a reasonable guess. By turning the volume up on the radio and comparing the audio Morse identifier with the code symbols illustrated on the chart he confirmed the station identification and hence, the frequency he had selected.

He had it made. He could select any frequency needed. All he had to do was count the clicks as he changed from the known frequency to any frequency he desired. "Now that I'm alone", he thought, "maybe I can make something of this day."

There was a clue to a possible answer to the question, "Why does Dave enjoy being alone?" Things seemed to operate better when there was no one attempting to impose their assistance when it wasn't invited.

Darkness was rapidly approaching, so he started making "blind" calls on the radio looking for an open airport. He had decided to continue the trip and was looking for fuel. Receiving no answers to any of his calls led him to think, "Gee, they really go to bed early around here." The obvious solution to his problem was to deviate from his route and land at Louisville.

When leaving Bowman Field at Louisville he told Louisville Departure Control that he was "squawking 1200 and enroute direct to Cincinatti." Louisville responded with the usual, "Radar contact, squawk **** for radar surveillance and vectors." "Negative on the vectors", Dave responded, "just letting you know who I am." (In a "Stage III Terminal Radar Service Area, it was not mandatory that you participate in its services in 1975.) More conversation with ATC followed. Very quickly ATC knew what was happening and understood his problems, which were a long way to go, a slow flying machine, deteriorating weather, and generally a long evening ahead of him.

As he flew along he heard ATC occasionally talking to other aircraft and eventually, in a conversation with a south bound twin-engine airplane he mentioned Dave's flight. "There's traffic at twelve o'clock, three thousand." He told the pilot of the twin. (An interpretation would be, Dave was directly in front of the twin at an altitude of three thousand feet.)

"No contact." said the pilot. (He didn't see Dave.)

"It's a sign headed zero four zero."

"A what", asked the pilot?

Knowing that the pilot of the twin couldn't see his marker beacon and position lights in the haze ahead and ground clutter below him Dave turned the sign on.

"Wow, I see it." The pilot announced. "Whatcha gonna do with that?"

"Take it to Detroit." Dave responded.

"Why?"

"A friend of mine wants to get their money with it."

The twin pilot's response was, "I just came from there and they ain't got any."

Dave turned the sign off and continued on his way in silence. The course took him by Cincinatti and he was "handed off" from Louisville RAPCON to Cincinatti RAPCON about halfway between the two cities. (RAPCON is the acronym for, Radar Approach Control.)

(Most ARTC facilities in 1975 were equipped with a radar system known as ARTS III. ARTC is the pseudonym for Air Route Traffic Control and an ARTS III system can show the controller the ground speed of aircraft equipped with transponders.) Sometime after he was handed off to Cincinatti RAPCON he was asked, "What kind of aircraft is that?"

"A Cessna 150 with a sign attached."

"Can't you carry it inside?"

"Wish I could. What do you show on ground speed?"

"Fifty five."

Shortly after that contact he noticed the reply light on the transponder had stopped blinking, so he played with the intensity control and it still didn't function. "Do you still show Cessna 18623?" he asked.

"I think I just lost your transponder. I show a primary target that may be you."

"O.K.. I indicate a transponder loss too." Dave responded.

"Where'd you come from?"

"Nashville."

"How's fuel?"

"O.K. I stopped in Louisville."

"Where're you goin'?"

"Detroit."

"You ain't gonna make it at that speed." The controller commented.

Dave was wondering what else could go wrong on this trip. The stop at Dayton for fuel and a candy bar was totally uneventful and he departed at midnight. As he approached Toledo he repeated the fact that the transponder had failed and asked,

"Can you call Detroit and get a clearance to enter the TCA for me?" (Detroit was known as a STAGE II Terminal Control Area. A transponder was one of the essential pieces of equipment required to fly in such an area. FAA would allow an operation within such an area once in such cases as equipment malfunction.)

Toledo RAPCON did call Detroit RAPCON and advised them of his problems and while relaying the clearance asked, "What's your ground speed?"

"Fifty five."

"Why so slow?"

"I'm carrying a sign."

"A what?"

"A sign made of a lot of lights."

"Wish we could see it". was his comment.

Dave turned the sign on, banked the airplane to its right, and said, "Here it is. May be too far out for you, though." He estimated that he was about fifteen miles east of the airport.

"I think we see it."

Fifteen minutes later Toledo RAPCON said, "Contact Detroit now on 125.15 and have a nice evening."

"Thanks much. So long."

The people in ATC had consistently been very helpful and cordial throughout the trip. That was not to change.

"Good evening, Detroit. Cessna 18623 with you."

"Good evening Cessna with a sign. Alpha is current and understand your transponder is out." (By referring to "Alpha" the controller was talking about the ATIS, the Airport Terminal Information Service that is a recording that is broadcast so that pilots may learn what runway is in use, which way the wind is blowing, the setting for the altimeter, and any other important information to them concerning the airport of intended landing. This relieves the controller from tying up the air traffic control frequency with this information for every airplane that approaches.)

"Affirm".

At three o'clock in the morning our nation's airways are not busy. There is very little conversation on the airwaves, so after a long silence Dave said, "You can't give me a ground speed, can you?"

Dave knew that with the transponder off line the controller could not provide such information, but perhaps he just wanted to hear another human being and after what he had experienced in the preparation for and execution of this flight he may have just wanted to hear the reassurance of a professional.

"No, but you've been with me for fifteen minutes and gone fourteen miles."

'Thanks."

"Cessna 18632, contact tower now, 121.1 and have a nice evening."

"Thanks, you too."

The tower's response to his call was, "Cessna with a sign, good evening, cleared to land. What runway do you want?"

"Three center. Going to Page." ("Three" is the runway identification number. Page was the company that provided fuel and other services at Detroit Metro and was located in the original Michigan Air National Guard hangar on the east side of the field.)

"Cleared to land three center and what's the sign look like?"

"I'll turn it on and make a low pass followed by a right pattern for three center if that's O.K." Dave replied. "You can take a look."

The tower responded with, "We show you five out. Cleared for low approach, right traffic, and landing. Turn on the sign."

Dave executed those directions and while the airplane was being refueled called Bill by telephone.

They put the little airplane in the hangar, went home, and returned to their routine schedules. Several thoughts about the trip presented themselves from time to time after its completion.

The first is, Dave knew that he enjoyed it despite the lack of communication from the sign manufacturer. The second is, He never enjoyed listening to the criticism of controllers from people who didn't comprehend controller's duties, responsibilities, and liabilities. The third is, He was never comfortable around people too proud to admit they had a problem in their jobs that consequently became detrimental to their performance.

The fourth is, He wonders if his love of flying is really an attempt to escape from the little annoying behavioral characteristics that persistently exhibit themselves in society. He was sure that he loved flying far more than he did society.

Two days later he walked into the office of Page Airways and one of their employees approached with a smile and said, "You should have seen Jerry yesterday. He had really been had by the tower."

"Oh?"

"Yeah. Night before last someone came in here with a weird arrangement of lights hanging on an airplane. Well, the tower had the pilot turn on the lights while he was still some distance south west of the airport. Then they called over here on the phone and asked us to confirm the sighting of a UFO to the southwest. It looked strange, the way the lights were flashing from left to right in repeated sequences on the thing."

"Well, Jerry ran back in the office, grabbed his new camera, ran back outside, and started taking photos of it. When it became obvious that it was coming to the airport he really got scared. Turned out to be some sort of a sign to be used at night."

PART XXXV

THE INTERNATIONAL AEROBATIC CLUB, CHAPTER 88

Bill Unger had his Private License and had begun the construction of a Starduster II biplane in his family room. Kim Kovach was flying his father's Citabria, and Jay Cavender was flying the Luscombe that his father had bought for him. Mark Johnson was flying a new Citabria that his father had purchased for him and all of them were mildly interested in a new sub-division of EAA called, the International Aerobatic Club. The identity of the person who originally suggested the attempt to organize an IAC Chapter in the Detroit area was undoubtedly either Kim or Jay, but among those three being led by the elder, Bill Unger the proposal was made and flourished if you can call the number, thirteen a flourishing group. There were a total of thirteen people listed as members on the Charter that was issued by IAC Headquarters in 1973.

Sanctioned and organized aerobatics in the United States originated in the early 1960's after the Texas crop duster, Frank Price went to Balboa, Spain in 1960 financed out of his own pocket and entered the World Aerobatic Championship Contest flying his modified Great Lakes 2T-1 with a Warner radial engine on it. Of course Frank Price and the antique airplane were not competitive, but he did open a lot of eyes and the stories about his participation in that first modern "World Aerobatic Championship Contest" are legion. Mr. Price returned with a copy of the, "Aresti Catalogue" of aerobatic maneuvers carefully and fully listing the maneuvers with assigned, "degrees of difficulty" based on the author's (Jose Aresti) experience with his Bucker-Jungmeister biplane. Prior to that time Champions in the United States had been named by the press' observation of what was essentially an air show in Miami, Florida.

After returning to this country Mr. Price organized his, "American Tiger Club" which sponsored a picnic and aerobatic contest at his home base, Waco, Texas and in 1964 the Aerobatic Club of America (ACA) was formed and began to have an annual contest to name the United States Aerobatic Champion. Federation Aeronautic Internationale-CIVA awarded World sanction to ACA and the seed of the sport was planted in the U.S.

By 1970 the group of known aerobatic pilots in the U.S. had grown to a significant number and there had been several airplanes capable of aerobatics conceived and born under the auspices of the EAA. Several pilots were becoming less than appreciative of, "Pappy Spinks" and the manner in which he was directing and influencing the ACA and the pressure of the internal disagreements was felt in EAA and by 1971 the IAC was born as a Division of EAA.

IAC used the same method for determining Chapter identification numbers as EAA had been using since its birth namely, pick the next number in the progression and that's your Chapter I.D. The pilots in the Detroit area are as determined to express themselves as any others in the country, so they selected a higher number that would be acceptable to IAC Headquarters in Oshkosh. The number 88 is the same whether being upright or inverted, so it was a logical choice for the Chapter I.D. and IAC-88 was born in 1973 with Bill Unger as the first President. The first meeting was held in the Quonset hut that was the home for EAA, Chapter 113 and sanctioned and organized aerobatics had come to Michigan.

Sanctioned aerobatics had several participants from Michigan prior to that time, but they were never known or acknowledged outside their own fraternal group. There were names like Art Davis, Bob Lyjak, and Bill Barber, but no one knew that Bob Lyjak had been very competitive in Aerobatic Contests sponsored every year by the Antique Airplane Association and Bill Barber had been named Captain of the U.S. World Aerobatic Team that went to Budapest, Hungary in 1962. Art Davis had been a nationally known air show pilot in his Waco in the 1930's and there was a newcomer to the sport, Henry Haigh who was competing in the Aerobatic Club of America Advanced Class.

The first IAC sanctioned event sponsored by Chapter 88 was an, "Achievement Awards" event that took place on Sunday, June 8, 1975 at Carl's Airport located south of Detroit with seven pilots participating in two Pitts biplanes, three Citabria airplanes, one clipped wing Cub, and one Stampe biplane. Dave was awarded the Sportsman and Intermediate Category Achievement Award. They're called, "smooth badges."

The "Achievement Awards" event was simply flying under the observation of three Aerobatic Judges that had earned the certification from IAC headquarters. Dave was the chairman and organizer of the event and was assisted by Clarence Landowski in the planning for it. The event also made a favorable impression on FAA Inspector Carl Borchers who was the FAA Monitor for the event and the Inspector assigned to direct the FAA "Volunteer Safety Counselors", one of which was Dave.

The "hit" of the day was the presentation of a birthday cake to Bill Unger by his wife, Bonnie. Bill attempted to blow out the candles on it and they kept relighting themselves even after he used a fire extinguisher on them. He finally gave up on the candles and attempted to slice the cake, but it also exhibited immunity to that attack. The firemen who were attending the event handed him a chain saw from the fire truck and it wouldn't cut it. Of course everyone was laughing too much to eat the cake by then. Investigation disclosed that the cake was only icing spread over a block of wood.

"Doc" Kovach loaned Dave his Citabria and he flew his first contest at Medina, Ohio on July 11, 12, & 13, 1975 competing in the Sportsman category. He won the "Grogan Belt" for finishing in the last position because he had flown across the "deadline", a line between the aerobatic area and the spectators where flight is forbidden. The rules specified that a "0" score for the entire flight be issued in that case. In 1976 Dave was there again in "Doc's" Citabria and finished in the middle of the group flying "Sportsman". Sherry went with him to that contest. He didn't attempt another contest until June 12, 1976 when he and John Maxfield flew Doc's Citabria to Salem, Illinois. He expected to do well in that contest until Giles Henderson landed in his Clipped Wing Cub. Giles

Henderson was probably the greatest Clipped Cub pilot that ever lived. Then on June 17 he and Kevin, Kim's younger brother flew the Citabria to Atlanta. The weather was poor and they flew only two flights. One was practice and the other was for competition. He forgot to perform the loop in the sequence and in disgust with his performance he left the contest without learning what position he had finished in. On July 16 he flew "Doc's" Citabria to Medina again and finished in the middle of the pack.

The first aerobatic contest that IAC-88 sponsored took place on the weekend of July 8-10, 1977 at Milan, Michigan. Dave was the, "Contest Director" and he did compete in the Decathlon that he was instructing in for Wolverine Aviation at Willow Run Airport. It was the first and only time he had ever flown a Decathlon solo and from the front seat. It was also the first, last, and only time he had flown a competition sequence in a Decathlon. He finished third in the "Sportsman" Category that had nine competitors in it. The contest only had the Sportsman, Intermediate, and Advanced Categories because the Unlimited Category Sanction was still issued by the ACA. IAC was only authorized to issue sanction for the first three Categories. The significant thing about that contest was that in Michigan aviation history it was the first aerobatic contest ever held in the state.

During the 1970's the listing of Presidents of IAC-88 was, Bill Unger, George Lytle, Jim Bole, Dennis Houdek, Art Patstone, and John Gardner. They were followed by Don MacDonald, Mary Lou Leverance doing the job for Al Wells, and Paula Elliot doing it for Len Rulason. In 1990 Dave McKenzie was given the office and held it through 1991 when Dick MacDonald was elected and held it until 2008.

John Gardner had been racing boats and then discovered airplanes in the early '70's, found a Pitts biplane that was for sale in Grand Rapids, Michigan, bought it and then ground looped it on the way home. After he brought the airplane home, had repairs made, and refused accepting the psychological pressure to exhibit his new skills and capabilities truly mastered the airplane.

John Gardner had established confidence in his ability to handle the little airplane that he based at Oakland-Orion Airport that was more commonly called, "Allen's" located on the east side of the city of Pontiac, Michigan. It was a grass field with two runways, the main one being 09-27 that was later paved. It became the site of the Chevrolet Assembly Plant in Orion Township, Michigan.

The fuel pump was located in front of the office and when John really began to feel comfortable in the airplane he couldn't resist exhibiting his confidence and after each second flight he would taxi briskly toward the fuel pump at the end of the day, pull the mixture control stopping the engine, turn the magneto switches off, release his seat and shoulder belts, and his parachute harness, and stand in the seat reaching for the hand hold on the upper wing center section. Then he would step out of the cockpit and onto the lower wing and place his other foot on the ground at the moment the propeller stopped rotating and the airplane stopped moving. All of those operations were conducted in one smooth movement with great élan until the day that he missed the handhold on the upper wing and fell out of the airplane breaking his wrist. The pilotless airplane stopped at the pump and waited for refueling.

John went through several modifications and improvements on the airplane which included changing engines from the 125 H.P. Lycoming to the 150 H.P engine, then to the 180 H.P., changed the original wings from the single aileron, "flat" wings to a set of double aileron "round" wings that he purchased from Henry Haigh. Then with the assistance of Bud Drum he installed the six cylinder, 540 cubic inch Lycoming. He was among the first in the World to fly that iteration of the Pitts biplane that began its reign over American Aerobatics with Betty Skelton's, "Little Stinker"

John Gardner had truly become an aerobatic pilot contending for a position on the U.S. World Team by 1986, but his aerobatic career ended when his wife, Paige's objections finally had an effect. The last heard from Paige and John was that they were living in Muskegon, Michigan and were both happy with boating on Lake Michigan.

Aerobatics is not a forgiving sport nor is it free of hazards and penalties requiring "paymentof the dues". It is not inherently a dangerous sport, but is terribly intolerant of recklessness, carelessness, and a lack of respect. There were casualties in IAC-88, perhaps more than in any other Chapter of IAC. The Chapter's first loss was F. DeWitt Barnard, the second was Steve Van Eyk, third was Harold Chappel, Jr., and fourth was Sam Gorsin.

The second through fifth sanctioned contests in Michigan were held in Owosso in 1978 through '81. The contest in 1979 was the first in Michigan to have the "Unlimited" Category. The next three were held in Sandusky, the next twelve held in Marlette, the next one in LaPeer, and the rest of them in Jackson at the towered airport, "Reynolds Field".

During every contest except the first one at Milan there had been at least one airspace intrusion. The FAA cited, "Safety" as the reason to prohibit aerobatics in positively controlled airspace and it was only after moving to an airport with Type "D" airspace that the potentially hazardous intrusions ceased. The FAA should revise Federal Air Regulations, Part 91 to allow aerobatics within Type "D" airspace at anytime an aerobatic pilot requests it with exceptions made for other reasons like conflicting traffic, of course. The FAA consistently exhibits a disregard for the safety of aerobatic aircraft and pilots by requiring that they practice in Type "E" airspace and simultaneously issues clearances for other traffic to fly direct to a fix or a destination with no concern shown for VFR traffic. Of course, VFR traffic is expected to use the, "See and be Seen" rules and when operating in VFR conditions "instrument" traffic is expected do the same. How can the aerobatic pilot feel secure when ATC may be vectoring who knows what into him at anytime?

The FAA has a great responsibility attempting to keep aviation safe and to protect the public. That's the justification for the public to financially support the FAA.

It is inevitable that something new will invariably appear, sooner or later. At the aerobatic contests it is essential that it be possible to bring a competitor's flight to a halt for safety reasons. Usually it's due to a traffic incursion, but there can be any reason for a "recall signal" being used such as the competitor flying lower than allowed, too close to the spectators, perhaps a part came off the airplane and the pilot is not aware of it, or an infinite number of other reasons.

Prior to the requirement in the IAC rules that competitor's airplanes must be radio equipped everyone involved had to depend on the "recall panels" and the "recall signal", both of which are under the control and command of the Chief Judge who really runs a contest. At Owosso in '79 during one of the flights the usual intruder appeared. Intruders were usually someone who heard of the contest and simply wanted to fly in and observe. Their standard explanation for intruding was, "The briefer at "Flight Service" didn't mention anything about an aerobatic contest." The FAA had insisted that the contest sponsor, IAC-88 have an ambulance, a fire truck, and smoke bombs to use for recalls.

When the intruder appeared and the "recall panels" were scattered to recall the pilot in the air at the time the FAA Monitor demanded that a smoke bomb also be released as additional notice to the competitor to break off his aerobatic sequence and land the airplane to have the "Starter" explain the problem. The smoke bomb set the grass on the airport afire and the fire truck was put to use to save the airport. That was probably the only time that a real emergency ever occurred at any of the Michigan contests and it is hoped that IAC-88 will be alive and well for a long time to come.

PART XXXVI

THE ULTIMATE DREAM

By late 1976 the Great Lakes replica was complete. Very few of the parts had been cleaned or primed and none had been painted, but all of the parts of the airplane existed.

John Maxfield seemed to be spending four evenings per week from 8:00 P.M. to 10:00 or 10:30 assisting Dave in removing, inspecting, packing, or loading them to go to the plating shop or sandblaster, and perform the final finishing, painting, and assembly.

A registration is required. It is law. Why, though, accept any number the beaurocracy might issue? In honor of the International Aerobatic Club, Chapter that he was so active in Dave chose the number 88. The "N" was, of course the letter that indicated that the airplane was registered in the United States and the letters, "SK" were his wife's initials. And so, N-88-SK came into existence.

During the winter of '76-'77, and the spring and summer of '77 they completed the covering and painting of N-88-SK with Kim Kovach spraying the final color coats on the wheel pants, landing gear fairings, all of the tail surfaces, wings, and struts. Dave did all of the masking and other preparations with John assisting on the sanding, hanging of parts for painting, moving into storage in the basement of the house, and too many other things to possibly remember, but they were essential parts of the job.

On the first Sunday of October 1977 some components of the airplane were moved out of their six and a half years of residence in the "airplane factory" and others were moved out of the ten year residence in the basement of 26034 Woodbine Drive in Inkster, Michigan. The wings were loaded in the school bus used as a camper by Howard Gilmore. The empennage surfaces went in the van belonging to Don Smith and the fuselage with the engine attached was loaded on an "A" frame made of 2x12's eighteen feet long and attached to a junked truck axle that belonged to Stan Wallis. John's station wagon carried ailerons while Dave's carried tools, struts, and anything else that lay loose.

Westward on Michigan Avenue out of Inkster, through Wayne, Westland, and Canton the caravan moved, its progress marked by turning heads and erratic paths steered by other motorists. After approximately forty minutes the caravan turned into the driveway at Mettetal Airport in Canton where it was greeted by a migration of the, Airport Bums Club from the office to the hangar rented by Stan Wallis that would be its home for the next two months.

Ruth Gilmore and her fiancé's (Paul Murphy) sister brought out fried chicken, potato salad, and other goodies for the consumption and pleasure of the group as the hangar was cleared out and parts of N-88-SK were moved in. Stan Wallis used the "assembled trailer" to take the fuselage of his, "Red Wing Black Bird" home to make some modifications to it.

Within an hour of arriving at the airport assembly began by attaching the struts for the center section of the upper wing and then the center section. The lower wings, interplane struts, and upper wings were loosely installed and then the recognition and acceptance of fatigue arrived coincident with the end of the day.

The completion of the airplane then dominated his life because he wanted to complete the assembly and fly it prior to the arrival of Michigan's winter which could arrive in as soon as four weeks. The hangar rented by Stan and loaned to Dave had a dirt floor, no interior finish, no insulation, no heat, and ONE 60 watt light bulb in the roof that was on the end of an extension cord. The folding doors were pretty good ones though.

Time presented the major challenge to Dave at this point. His efforts to solve the problem consisted of ceasing to teach students and his daily schedule became going to the office and remaining all day, leaving the office and going home at 4:45 in the afternoon, eating dinner, changing clothes and driving to the airport, and about 6:30 in the evening begining work on the airplane.

John Maxfield was attending college every morning and then working for Timoszik Aviation at the Ann Arbor Airport until 7:00 and then would arrive at Stan's hangar to assist Dave around 8:00 o'clock in the evening.

Dave lost track of Mike and Terri's activities during this period and on November 2 the fourteenth anniversary of Sherry and Dave's wedding arrived. It was also Sherry's birthday. Oh, how well he remembered that anniversary and birthday. Sherry didn't insist on dinner or anything traditionally associated with those celebrations, but she went to the airport and assisted Dave working on the Great Lakes. Was there any love in that union? Yes.

The FAA Inspector had to see the airplane twice during that period. The first time with the airplane assembled, but "opened up" so that he could confirm continuity of controls, security of fuel lines, oil and hydraulic lines, wiring, etc. The Inspector performing that inspection was Ben Rowland from the Detroit office and he was known for his demeanor. He never exhibited a smile, never said hello, never asked how you're doing, and Dave had never heard him say, "Goodbye". When Dave opened the hangar door Mr. Rowland asked, "What is this?"

Dave's response was, "It's an airplane".

Mr. Rowland was not amused. He inspected the airplane and then said, "O.K., close it up and then call me again. I want to see it ready to fly."

It only required a couple of nights for Dave to "close it up", call Mr. Rowland, and prepare for his final inspection.

When Mr. Rowland arrived he had another Inspector with him and Dave had the tail wheel of the airplane tied to his Pinto car. Mr. Rowland asked, "What's that about?"

Dave responded, "Well, this airplane does not have a starter on it, so when you're ready to hear it run you'll have to sit in it to handle the engine controls while I hand prop it. You may like the airplane and I don't want you to run away with it."

Mr. Rowland only said, "I'll type up the airworthiness certificate and put it in the mail this afternoon." Dave then heard him very quietly say to the other inspector, "Did a pretty good job, didn't he?" Dave obviously wasn't supposed to hear that.

The airplane that the FAA had decided to list on its aircraft registry as the, "McKenzie-Lakes" had passed its weight and balance checks, its five hours of ground run time on the engine, twelve runs down the runway with the tail wheel in the air, the inspections by the FAA Inspector, and Dave's own inspections. It was ready to fly.

Dave telephoned his parents in Alabama and advised them that the weekend of November 19 and 20 would be the targeted time for the first flight and they promised to be in attendance at the event with his brother, Jimmy.

While awaiting their arrival Sherry, Dave, and the Maxfield family, Virginia, Orlo, and John attended a showing of the movie, "Bandit". Sherry, being the "deep water, born again" Baptist had only been to two movies since seeing, "Goldfinger" in 1964. One was, "Those Magnificent Men in Their Flying Machines". The other one was, "Midway" and it was only because Dave's Aunt, the wife of his favorite Uncle from Augusta was visiting them.

It is a good policy to have a competent physician at the scene when one is performing a potentially hazardous activity and with that in mind Dave had Told "Doc" Kovach that November 19 would be, "THE DAY" and he would telephone him with more details as soon as they were known.

The airworthiness certificate arrived in the mail on Saturday morning and the weather was perfect for a test flight. The weather forecast was predicting the passage of a cold front Saturday evening through Sunday morning, so after eleven years of effort and $12,000.00 dollars in expenditures, the inconveniences, the bruises, scratches, tolerance of critical comments, embarrassment, and pessimistic predictions Saturday afternoon was going to be, "THE DAY" he would fly the "Lakes". He still regrets that he overlooked the call to "Doc" that afternoon.

By 2:00 P.M. Saturday afternoon he could see the cloud line preceding the approaching cold front and he knew the airplane was ready. He had maintained his personal proficiency as a pilot over the last ten years as a free-lance flight instructor and he had flown two other Great Lakes airplanes. The first was a homebuilt replica with a

Ranger engine on it in Louisville, Kentucky that he flew in '72 and the second was the "new" factory prototype from Enid, Oklahoma at Oshkosh in '75. He knew what to expect.

He had become acquainted with a TV News reporter who was working for "Channel 2" in Detroit, Tom Korzinowski who had started flying. Tom and Dave would see each other at Page Airways on Detroit Metropolitan Airport where Dave was an instructor and Tom was a student. Tom had given Dave the telephone number for Channel 2's news assignment desk and asked Dave to call if he were going to fly the "bird" on Saturday. Tom was going to be in Lansing, the state capitol on Saturday morning, but if he was back in the station he wanted to bring a cameraman and witness the test flight. Dave had taxied the airplane from Stan's hangar to the fuel pump and made sure that he had twelve gallons of fuel aboard for the flight and then made the telephone call.

"No, Tom is still in Lansing." was the answer he received. Then the voice on the line continued, "Tom has told us about this, so give this number to someone else and tell them to call us if it doesn't go as hoped for." Dave hung up the telephone and has not to this day watched Channel 2 in Detroit.

Pete Prince in Milton, Florida, who had once worked for the Gee Bee factory located in Springfield, Massachusetts had worked on the wooden wings of all thirteen of the high performance Gee Bee racers built in the early thirties and had told a story about one of the airplanes having such a huge engine and small wings that they had put an offset in the throttle quadrant to prevent inadvertent full opening of the throttle on a take off before sufficient speed had been attained to control a "torque roll". He had gone on with the story saying that while enroute to the west coast for one of the races the pilot did lose his life in a half roll after take off and the investigation revealed that the throttle was at the full power position.

Dave's dad, Mac, having been a pilot during that period in aviation history reminded Dave of the story and Dave agreed to make one more taxi run down the runway to demonstrate the Lakes' capability to remain under control at full throttle and slow speed.

John Maxfield hand propped the airplane and after the pre-take off checks were completed Dave made the run at full throttle and immediately after closing the throttle wished that he had never attempted the run. He had to use extreme braking efforts to stop the airplane prior to reaching the end of the runway and just barely succeeded. He and Mac thought that he was going to lose it on that rollout.

He taxied back to the approach end of Runway 18, checked the trim setting, the alternate air selection, performed another magneto check, and TOOK OFF.

Following lift off he felt so good that he did a little wing waggle to let everyone know that it was all right. He climbed to about 2500 feet altitude AGL and explored normal and steep turns, cruised at 1650 RPM because of the low engine time, flew at minimum controllable airspeed, cycled the trim control and did a full set of stalls. Specifically, straight power off and on, approach, departure, and accelerated stalls, in that order for both right and left directions. It should be noted that he had pre-planned the flight and had a check list written out and on board. Then descended, entered the pattern for Runway 18 and made a low pass with a wing waggle for the cameras that several friends were operating, made a climbing turn to the cross wind leg and pattern altitude, completed the circuit and made a wheel landing.

He taxied back to the area where most of the friends and family were gathered and shut down. There was a brief silence, then applause, a kiss from Sherry, and the questions began. He knew then why no one knows what Charles Lindbergh's first words in Paris, France were. In such confusion and joy, who could care?

That evening Sherry presented a card with the inscription, "Why put off 'til tomorrow what you can do today? If you like it you can do it again."

The memorable experiences in an aviation oriented person's life might be the first solo, receipt of the Private Pilot's Certificate, their first instrument approach when they're the only pilot aboard, the first slow roll that comes out "nearly right", and perhaps being hired by an airline. Now, he had experienced the "Ultimate Dream". He had flown the airplane that was built by him. At least that's what he thought at the time.

Dave McKenzie

Great Lakes N-88-SK enroute to Grosse Ile, Michigan December 3, 1977 (photo from author's collection taken by Gail & Bob Jackson)

PART XXXVII

THE PRICE

When he was attending an early meeting with the people involved in the sport that he so dearly enjoyed he would always picture the man sitting on his front porch awaiting his arrival.

They had agreed to meet and go to Jackson to rent the Citabria, a Champion 7KCAB so that Dee could give him some dual instruction in aerobatics. The agreement had been made with very little conversation and, he supposed it was assumed that he would provide the transportation and pay for the airplane rental. There was no mention of compensation for Dee. No exchange of money was ever made between Dee and him.

They went to Jackson and flew two times. The first time they flew was on May 31, 1971 and the second on July 18, 1971. Dee taught him inverted flight, barrel rolls, slow rolls, how to perform immelmans and Cuban eights. He also taught him how to do point rolls, and probably most of all he helped Dave increase his own confidence.

At that time Dave took the friendship for granted and assumed that Dee did the same. Dee wouldn't have asked that it be done any other way. It was one of the few friendships Dave could remember where he truly received more than he gave.

They first encountered each other at a meeting of the EAA Chapter at Mettetal Airport. Dee didn't make a lot of noise. He didn't even seem to be involved in the operation of the club, but during the "coffee break" his depth of character became apparent. Dave knew that the little guy with the bowed legs and one-day-old beard was not a dreamer or pretentious person. Something about him said, "Experience, competence."

Over several years Dave learned that he held a Commercial Pilot's license and Flight Instructor's rating. He had been a fighter pilot during WWII and was stationed in Alaska in the Aleutian Islands. He was one of two people Dave met who flew the P-40. Remember the "Flying Tigers?" That's the airplane. His was called an, "Aleutian Tiger". He had owned a Great Lakes at one time, but when Dave became acquainted with him he owned a Ryan SCW. Wow, That's a rare airplane! He once complimented Dave by allowing him to fly it. Dee was with him, of course. Some years after that he complimented Dave even more by asking him to fly it back to Ann Arbor, Michigan from Oshkosh, Wisconsin solo. He needed to return to his job and the weather was too bad for him to fly it home, so he was going to leave it until further arrangements could be made.

He wondered why Dee would have had so much confidence in such a person as him.

Dee owned a shop that chrome plated parts for the auto industry and his acquaintances used to say, "He makes his living chrome plating pennies and passing them as dimes." He was never known to attempt a response to that allegation. He was just known as, "Dee, the pilot."

He eventually sold the Ryan and none of those who knew him could picture him without an airplane for long. He probably felt incomplete without one. It sure looked that way to his friends.

Dee, Arnold, and Ken formed a partnership with each other and decided to build an aerobatic airplane. They expressed interest in a "clipped wing" Taylorcraft and another acquaintance of Dave's had an old T'craft airframe. They thought this was a logical beginning for their airplane and bought the collection of parts. When they started rebuilding it they found that the fuselage had too much internal corrosion to be salvageable, whereupon they sold it to an aircraft salvage yard. (A "junk yard" if you prefer.) The operator of that yard displayed the fuselage as a "useable" aircraft component. The story caused Dave to be suspicious of the yard owner's principles and integrity and perhaps he just didn't know the fuselage was essentially junk.

Having given up the Taylorcraft project, they bought a partially constructed Stitts Playboy from Marion Cole. In a couple of years they had completed it and were flying.

Ken had an accident in it. Damage was minor. They had to remove and store the wings while they repaired the fuselage and landing gear, but being creative, and certainly competent people, they finally completed the repairs and were flying it again.

Dave watched Dee flying it in aerobatic contests in Ohio and Michigan and along with several others wondered how Dee could perform so well in such an airplane. Some even wondered whether or not a Playboy should be flown in aerobatics.

On the afternoon of August 13, 1979 Dee went flying in the Playboy and when darkness fell he still hadn't returned to the airport. The next day they learned that he had headed south at very low altitude over another friend's privately owned landing strip and pieces of the covering material on one wing started tearing away. Specifically, reinforcing tapes were tearing away. Dee ascended nearly vertically until the airplane changed directions of flight, descended vertically with the nose down and upon impact, it is said that the "engine went in five feet." It was instantly fatal to Dee.

It is suspected that while the wings were in storage in an outbuilding while repairs were made on the fuselage mice may have eaten away some of the rib stitching. That could have led to the failure of the wing covering.

Dave's family and he attended Dee's funeral. It was after the service was concluded and they were all standing outside the church that Dave turned away from everyone, hung his head, and wept. Dave, the restrained, unemotional, image conscious, wept in public.

That same day after taking his family home he went to the hangar, got the Great Lakes out and flew aerobatics for a while. It was the only way he could be totally alone. Even though he couldn't concentrate on the maneuvers and didn't care how precise they were that afternoon it was the only thing he felt like doing.

He thought Dee would have approved.

You know, he can still picture Dee sitting on his front porch awaiting his arrival exactly as he had on that day in May 1971. We pay a price for the pleasures we enjoy.

PART XXXVIII

CHANGES AND ADVANCEMENTS

In December 1978 the, "High Temperature Gas Turbine Engine" Department at Ford Motor Co. was closed. Dave had been in the department since July 1976 where they were attempting to develop a gas turbine engine with the "hot section" made of a ceramic material. Dave was transferred to Small and Mid-size Car Chassis Engineering that was supervised by Trion Moga, a supervisor that he had known and liked since they were both designers in Advanced Chassis in 1964. This was the type of work that Dave had joined Ford Motor Co. to perform. He had been transferred to "Vehicle Assembly Simplification and Serviceability Research" one month after the department was formed in the Scientific Research Laboratories and had worked in that department from December '69 until it was closed in the summer of '73. He was then transferred after an interview with his soon to be new management and their acceptance of his services.

There were approximately two and a half years spent in his new assignment and the time spent in "Alternate Engines Research" designing the installation of a Stirling engine in the '74 Torino car had been frustrating mainly because the management failed to consider the employee's interests, education, talent, and experience. There was also a personality conflict in the department that Dave was uncomfortable with and the whole problem was resolved when the company abandoned any idea of ever using the Stirling engine. There was always a feeling in the department that the company's upper management had no confidence in the concept and was spending the time, effort, and money solely to make the Environmental Protection Agency believe Ford was actively pursuing compliance with their directives.

In November 1975 he was told at 4:15 on a Friday afternoon to report to the Stratified Charge Engineering section of Advanced Engine in the Scientific Research Laboratories on Monday morning at 8:00 o'clock. That was it.

On Monday morning he sought out the supervisor of, Stratified Charge Engineering and was told, "Pick an empty board and move in. I have a meeting to attend and I'll see you later in the day." On Wednesday morning he asked the supervisor who the design leader was and what was expected of him?"

The reply was, "I have a meeting right now. I'll see you when I get back."

In July '76 Dave told the supervisor, "I'll be out on vacation all of next week. When I get back the following Monday morning I'd like to be told that another department needed my services and that I should report to, "fill in the blank". That was the extent of communication with the management in that department. He felt that they didn't need him, didn't want him, and damn well were not going to use him.

When he returned to the office from the vacation the supervisor said, "While you were out the High Temp Turbine Department started looking for another designer and you were selected. I'll be back by 10:00 and we'll go to their office for an introduction." He left the room.

AT 10:45 Tony Paluzny and Richard Jerrian from High Temperature Gas Turbine Engineering appeared at the door and asked, "Is there a Dave McKenzie in this room?"

Dave exchanged introductions and one of them said, "Come with us."

Dave enjoyed the people, the work environment, the supervision, the department in general, but there was no future for the department. There were serious deficiencies in the components called regenerator cores and the production of ceramic power turbines and stator vanes was extremely unreliable even though some had been produced that functioned acceptably. Ford Motor Co. sold the technology that it had acquired in the years of research to a Japanese company. Dave regretted the closure of the Turbine Engineering department in December 1978, but he was really happy to return to automobiles that really ran.

He was doing well in his hobby, aviation. He finally had an airplane that fit the requirements of his desires and wishes. He enjoyed a very good reputation as a flight instructor. His son looked like him and was doing O.K. in school. Terri seemed to be doing just fine. He thought his marriage was strong and secure in spite of the vast differences in beliefs and standards of living that had become apparent between Sherry and he. He thought he had a safe, strong, and secure marriage and home. Their lives were organized, predictable, comfortable, and stable.

Sherry drove automobiles that were never more than three years old while Dave drove the ones that were three to six years old except for the period from '69 through '73 when he drove a street legal fiberglass bodied dune buggy built on a modified Volkswagen chassis that he had constructed for the family's second car. He was certain that Ford management really did not care for the dune buggy being in the employee's parking lot, but he was a free willed designer.

Yes, there were disappointments like the degeneration of the neighborhood that forced them to sell and leave the home that was close to Dave's office, small, but not too small, and had the heated "airplane factory" in the backyard. That was becoming obvious in early '77 before the Great Lakes was complete and John Maxfield's Pitts had been moved to the "factory".

When Mike was in the second grade a female student in the fifth grade acquired the assistance of two girlfriends and they physically assaulted and hospitalized the male math teacher for giving the instigator a failing grade on her report card. The Dearborn Heights, District 7 School Board and the School's Principal did nothing about the incident.

That incident inspired Sherry and Dave to enroll Mike in the private school system operated by a Baptist Church in the city of Allen Park. They enrolled Mike and Terri in Baptist Park Schools in the city of Taylor when Mike started the fourth grade and that would change about four years later. In the meantime they were paying school tuition for two children in a private school system and simultaneously paying the school taxes on their home and property. They felt that they were paying for four educations and barely getting two.

The house was not air-conditioned and they slept on summer evenings with the windows open. About 1:00 o'clock in the evening Sherry and Dave were awakened by the sound of gunshots that seemed to be in their backyard. Their bedroom was in the rear of the house and the shots were close. Dave called the Inkster police department. Dressed in his underwear, trousers, socks, and shoes he exited the back door that was on the side of the house and saw a police car turning into his driveway. He started to walk towards it and two policemen bounded out and lowered their pistols on him.

He quickly convinced the policemen that he was the caller and pointed out the lighted house behind his where there was obviously a party in progress. Soon there were five police cars and ten policemen in the neighborhood and they searched the house where the party was taking place. As they had approached the house they could see through the open windows that one person inside was wearing two shoulder holsters with pistols in them and two more pistols in his belt. When they rang the doorbell he promptly ran to a bedroom and attempted to hide the weapons in the closet, but they had already seen him and claimed to be in pursuit, so they didn't need a search warrant. The occupant of the house was arrested and the neighborhood was peaceful again. Dave signed the arrest warrant as the complaining witness.

About a month later the trial was on the agenda at the Inkster City Court and Dave had to leave the office to appear at the trial. The defendant and an attorney appeared when called, but the officer who had signed the documentation for the complaint and arrest was appearing in Wayne County court on another case, so the Judge dismissed the case because the arresting officer did not appear. The defendant was free from the Jackson Prison on probation for another offense, so Sherry and Dave decided that they had to move.

They were in a neighborhood where, "white flight" had appeared with a vengeance. Their property values were static and real estate values in other neighborhoods were increasing rapidly, so they listed the house with a real estate agent and started looking at other localities.

It was rather amusing when the realtor was conducting a tour of newly listed properties and brought a van loaded with agents to 26034 Woodbine Drive to show the property. Of course they found the basement full of wings and empennage surfaces painted in the red, white, and blue color scheme of the airplane and the "garage" with two airplanes in it. They had no idea how to evaluate it.

The property was sold in June of '78 and Sherry found a house in Wyandotte, Michigan that she dearly loved. Dave thought, "It's O.K.". They sold, purchased, and moved. Dave developed a hatred of the new home, probably because he had no place to pursue his hobby. Sherry and Dave discovered that the home on the south side was a duplex and the brick mansion on the north side was on the verge of becoming uninhabitable. When the garage in its backyard collapsed it fell onto the McKenzie's property. In addition to that the neighbors always parked in the street and it was impossible for two cars to pass. The solution to the problem was for one of the cars to turn into a driveway, let the other pass, and then back into the street and continue. It was a "blue collar" neighborhood.

One evening Sherry and Dave were awakened by a noise, but couldn't find anything wrong in the house and decided that it must have been a neighbor. The next morning Dave found that someone standing on the roof of the attached garage had damaged the window into the bathroom and when they jumped off the garage they had landed on the hood of Sherry's '78 Fairmont that was parked in the driveway and severely damaged it.

After six months residence in Wyandotte Dave said to Sherry, "I'm moving to Grosse Ile and you're welcome to come with me or you can stay here."

PART XXXIX

THE NEXT THREE YEARS

On December 3, 1977 Dave flew N-88-SK to the Grosse Ile Airport and became the first occupant of the new "T" hangar number 4 located on the old East Ramp. Prior to making that flight he visited Troy Ruttman's store at Mettetal Airport and purchased a snowmobile suit. He was attempting to prepare for a cold flight. It was four degrees above zero on the ground when he landed at Grosse Ile after having to fly west from Mettetal until passing Ann Arbor, then fly south about twenty miles and then turn east for another sixty miles. All of it was at an outside air temperature of approximately one degree above zero and in an open cockpit airplane with no heater. His logbook shows that the flight required one hour to complete.

He was followed on that flight by his student, Gail Jackson and her husband, Bob in their Cessna 172 to bring him back to Mettetal. He was then ready for the winter and he wanted to work on John's Pitts in the "factory" where the "Lakes" had been born.

John and he set the Pitts fuselage up in a level flight position, tack welded the engine mount together, modified the fuselage to attach the spring aluminum landing gear that John's dad, Orlo and Dave had given John for Christmas, built an aluminum turtle deck, and installed the empennage structure that they had obtained from Don Smith. Bruce Panzl came over and did the finish welding on the modifications and then a buyer for Dave's property appeared and John took the project to his parent's home in Dearborn and completed it.

By the end of June 1978 he had flown the Lakes fifty hours, the FAA required test program time, "opened" it up, and had the FAA perform their final inspection on it. On July 4 he made 3 flights in one hour carrying Sherri, Terri, and then Mike for their first rides in the airplane and on July 28 through August 1 he and Mike visited the annual EAA Convention at Oshkosh, Wisconsin in the airplane.

On August 8, 1978 he performed a demonstration flight for and received his first, "Statement of Acrobatic Competency" from FAA Inspector, Delbert C. Burgess and then could perform in public air shows with the airplane.

Mike had begun to show interest in the, "performing arts" and expressed some interest in becoming a radio announcer, actor, or journalist, so with his background of attending and observing various aerobatic contests and air shows with his father he thought that he might add, "Air Show Announcer" to his growing resume. He had been participating in the performing arts at Baptist Park Schools where he was a student.

On August 20, 1978 he made his first public appearance at Lamont's Airport near Deckerville, Michigan when his dad performed for a Fly-In of the Confederate Air Force, a gathering of World War II vintage military aircraft.

The next public appearance was at the Mid-Eastern Regional Fly-In (MERFI) at Marion, Ohio on September 9 & 10, 1978. Dave and Mike flew from Grosse Ile to Marion, Ohio in the airplane while Sherry and Terri made the trip in the Fairmont. An agreement had been made for him to fly in the air shows at the fly-in on Saturday and Sunday afternoons. On Saturday afternoon he was the third or fourth performer to fly, but on Sunday he was asked to be the opening act. The air show spectators must have enjoyed the performances of the Great Lakes.

PART XL

CAN THESE VEHICLES TRANSCEND TIME?

He only listened to the older man because of his age and out of common courtesy. After all, we are trained throughout our childhoods to respect and honor our elders. Even though the older man didn't appear or sound impressive or even interesting, it seemed proper to stop what he was doing for at least a short moment and allow him the satisfaction of some recognition and attention.

In most cases he would have demanded that anyone, even the elderly, remove their hands from the fragile and delicate surfaces of his machine, but in some way, for a fleeting fraction of an instant there was a faint spark of recognition between the older gentleman and Dave, so Dave didn't say anything as the gentleman caressed the curve of the wing tip bow and contour of the leading edge of the lower wing.

Instead, he walked around the wing tip and offered his hand as he introduced himself. It seemed that an introduction wasn't necessary, what with his name being painted on the side of the cockpit, but it did confirm his identification. It was also a continuation of the act of courtesy he had been taught to extend to people like the older gentleman. He may have mumbled his name. It might have been a complex and completely unfamiliar name or it may have blown away in the slight breeze. Maybe it wasn't meant for Dave to know his name. It didn't matter, anyway.

You do the restoration, son?

"No sir. It's not a restoration. It's a homebuilt." Dave replied as he wondered, "How many times per weekend do I have to say that?"

"From a kit?"

"No sir, from scratch. Every part came off the U.P.S. truck and through the front door as either a straight tube or a flat sheet." Replied Dave as the thought crossed his mind to install a tape player with all those answers repeated on it. He also felt himself bracing for the next question.

"Well, it sure looks familiar. It looks like a, a, a, - - -?"

"Great Lakes, modified considerably." was the voluntary response he made. Once more, out of courtesy he found himself rescuing the older gentleman from the corner he had talked himself into.

"Oh yeah, that's right. Fine flying airplane. Built up in Saginaw as a navy trainer, weren't they?"

"No sir, Cleveland, 1929 through 32, as just another biplane for fun flying." Over, and over, and over again every weekend, he answered the same questions with the same answers in the same order. They didn't even have to move their lips. He could see it in their eyes. The only identifying difference between them was a slightly different tone of voice or accent.

As was often the case he realized that he was on the verge of boredom and in an effort to avoid that miserable waste he asked, "You fly?"

"Used to."

And gradually the old fellow started to stand a little taller and straighter. The studied, deliberate attempt to maintain an expression of indifference on his face melted away as he said, "I flew back during the '30's and '40's, built two homebuilts." and then the old timer had presented his credentials, been recognized, was accepted, and had become a part of the fly-in.

In the background Dave noticed for the first time, the guy, fortiesh, the woman, and the two children. It didn't require a psychic to identify them as grandchildren, daughter, and son-in-law.

He readily saw that the guy would rather have been at home with a beer watching the Tigers loose to somebody or other. The children were for that moment interested in the airplane parked next to his and he knew that wouldn't last long. The woman? The woman was not so easily categorized. She had been relieved to see that someone had accepted her father and she was also concerned that he might bore Dave. She need not have been concerned.

"What sort of homebuilts did you build?"

"One was a monoplane similar to a Baby Ace. Used a Continental A-40 for power. Flew real well. A windstorm got it though. Kids came along wife never cared much for my flying, anyway. So, never got around to rebuilding it."

"Yes sir, I understand that well enough." Oh, how well he understood all of it. "What was the second one?"

"It was an EAA biplane. I didn't finish it. Sold it before it was covered and I'm sort of expecting it to show up here today." The gentleman said.

"Fine, now that your kids are grown, have you ever considered getting back into aviation a little more seriously?" Dave asked.

The elder gentleman was completely at ease by then and his concern for possibly disturbing Dave had been replaced by a small degree of amusement with Dave, the youngster, the upstart.

"Oh, I haven't gotten out or quit." The weight of years and responsibilities met and carried honorably rapidly left his being.

"I have a Kinner Bird that I'm rebuilding right now."

In that statement time warped. He became young and Dave grew old. The older man was then experiencing youth as young as the year of the vintage of the Kinner Bird." In that same instant Dave was overwhelmed with memories of his youth. Experiences and memories came flooding back and caused him to feel as though he was drowning. Suddenly he felt older and a little tired. The age marks on the old man's face and in his posture disappeared and then they were no longer an old man and a young man casually meeting on a summer afternoon. They were equals in all respects. They were pilots, mechanics, and builders of time machines.

The older man was looking into Dave's eyes for a hint of recognition of the name, "Kinner Bird", as Dave said, "Kinner Bird? My father had one back in the 40's. Owned it twice as a matter of fact." Memories and history were completely engulfing him then.

"Where is or was your dad?" he asked Dave.

"Down in Alabama. The second time he sold it, it was to a student pilot who mismanaged the fuel supply and had a forced landing in a field that also contained a ditch."

The older man waited to hear more.

"Another guy bought the wreckage and rebuilt it with farm machinery hardware and so forth, then wanted Dad to teach him to fly it. Dad wouldn't, so it sat on the airport in the weather for about two years deteriorating."

"What ever became of it?"

"Well, an ag pilot from Monroeville, Alabama bought it and that's the last I really know of it, for sure."

Dave knew he wanted to hear more by observing the manner in which he was smiling.

"Oh, ten years or so ago a gentleman from Ohio went through my home town and he had the paper work on the Bird. Wanted to have my dad relinquish title to it. Seems that none of the later owners had ever updated the title and it was still listed in dad's name by the FAA. Dad checked with Carter, the ag pilot and he confirmed the fact that he had sold the remains to the Ohioan. Dad said he had sold the airplane in good faith and had no real claim to it, so he signed a bill of sale for the gentleman."

"That's my Bird."

Dave didn't have to hear him say it. He already knew. With that a mutual bond was born without man-made agreement or arrangement. There were no longer any differences in age, no differences in experience, no differences in hopes.

They had both flown, then, now, and in the future. What one had done, or does, the other had done, does, and will do.

Then the loud speakers on the field interrupted with, "All air show pilots, please report to the terminal office for pre-show briefing with the FAA." Over and over that sound was repeated and echoed across the field.

"Oh, shucks, that's for me. 'Gotta go." Dave said. "Ya'll gonna' be around for the aerobatics?"

"Don't think so." the older guy said, "Daughter's got sompin' cookin'." Wants to get home."

The Spirit's Journey

Time warped again. He was again the unobtrusive little old gentleman bearing the responsibilities of life and family. There he stood again, the visitor to the fly-in. A little old man Dave expected to hear the same tired old questions and comments from.

"I'll tell dad we met."

"Wish him well for me."

"Yes sir."

The elderly gentleman was again the little old man as he rejoined his family and as Dave started walking towards the terminal thoughts surrounded him, isolated him from the gathering air show spectators. They were both spitits riding in bodies. The older man's spirit had been in his body a little longer than Dave's spirit had been in his, but they were the same age with the same experiences and same ambitions. They were each living in their own times and both of them were aware that their own times were occasionally warped by time machines called, "bi-planes".

The judges who inspected the airplanes named the Great Lakes as the, "Outstanding Homebuilt" at the fly-in and Terri rode in the front seat of the airplane to return to Grosse Ile. Mike rode home with his mother in the Fairmont.

PART XLI

LIFE DOESN'T CHANGE

Life goes on. Dave had the usual five students in the flight training curriculum one of who was Dr. James Fordyce whose specialty was allergies. Jim as Dave knew and addressed him had joined one of the flying clubs based at Detroit Metropolitan Airport and his search for an instructor led him to Dave. Jim Fordyce and his son, Jimmy were two of the people who had assisted in moving the components of the Great Lakes from Dave's home in Inkster to Mettetal Airport in the fall of 1977. Dave trained him from the start of his flying career through the Private License and the instrument rating.

Tony Malizia was employed at Ford Motor Co. when he sought out Dave to teach him to fly. Tony earned the Private License and the instrument rating with Dave. He left Ford and set up his own law office. Four times in later years Dave was very pleased to have Tony as a friend.

Gary Clark was a High School teacher who belonged to the same flying club that Jim and Tony belonged to at Detroit Metropolitan Airport. Gary earned his instrument rating with Dave and remained a friend after completion of the training for it.

Bill Appleberry joined the same flying club that Bill Unger had belonged to, MANG. Dave trained Bill for his Private License in the Cessna 150 and 172. Bill's wife, Marie and Dave enjoyed the telephone calls that he occasionally had to make to their home while Bill was one of his students. Both Marie and Bill are still friends of Dave's.

Mark Matusiak is the son of "Del" and John who have been friends of Dave's since their meeting in 1966. When Mark reached 16 his father asked Dave to teach him to fly. That was begun at Detroit Metropolitan in a Cessna 172 belonging to a corporate pilot that Dave knew. Mark's training had to be moved to Grosse Ile Flight Service when the corporate pilot moved to Reno, Nevada, but Dave continued to be his instructor and Mark completed the training for his Private License successfully. Mark went on to college at Western Michigan University, graduated, joined American Airlines, and became a Captain. To meet his military obligation he joined the Michigan Air National Guard where he became a member of the Air Guard Jet Demonstration Team flying the F-16.

On September 29 Dave flew the entertainment act at an annual open house at the home of Jan and Joe Rayne in Clinton. Of course it required the approval of the FAA, Flight Standards District Office that required the usual submission of the FAA form 7711-1, Application for Waiver and receipt of the approval for the operation, FAA form 7711-2.

Dave flew the performance for no financial compensation, just the acquisition of low altitude exhibition experience.

In October '78 his mother, father, and brother made a visit to Dave's "new" home on Grosse Ile and during that visit "Mac" flew the Lakes with Dave in the front seat of course. Then Dave changed seats and took his brother, Jimmy for a flight. He considered it important for those who had made significant contributions to the successful construction of the airplane to receive a ride in it.

Sometimes Dave would ask himself, "Why can't I be successful?" and life continued onward as he continued to teach aerobatics in a Bellanca 8KCAB based at Willow-Run Airport and conducting Biennial Flight Revues in many different airplanes.

PART XLII

THEY HAD AN AIR SHOW AT HOWELL

On June 16, 1979 the Michigan Chapters of the EAA sponsored a fly-in at Howell, Michigan. Doug Robertson, a long time member of EAA Chapter 13 was the head of the organization and should be thanked for trying so diligently for so long to make it work. Aerobatic performances were requested of the International Aerobatic Club, Chapter 88. Of course the highlight of the show was presented by Henry Haigh flying the "Super Pitts" that he had flown at the 1996 World Aerobatic Championships Contest in Kiev, U.S.S.R. Bob Barden, and John Gardner also flew aerobatic performances and Dave flew one in the Great Lakes and later said," It isn't all fun."

He had completed the entire aerobatic routine except the last three maneuvers and was entering the downwind leg of the traffic pattern thinking of the airspeed and reduced the throttle setting while slowing the airplane for the double snap roll. Suddenly he stood hard on the right rudder pedal and followed with a very positive pull on the stick he held in his right hand and then pulled the stick to the right. The airplane yawed to the right, started to raise its nose, and then the airplane, confused by the commands from its pilot, gave up flight and entered the spin in the horizontal direction. Around they went, the airplane and the pilot, once, then to one and a half turns, then he pushed the stick forward while he simultaneously added full throttle and changed

The Spirit's Journey

from right to left full rudder application. The airplane slowed its rate of roll and stopped rolling with the wings banked about sixty degrees to the left that he held as he made a turn to the left onto the base leg of the pattern.

The almost double snap was complete. He was on the base leg of the pattern and then he repeated the sequence of movements, but only rolled one revolution, stopped the roll with sixty degrees of bank to the left remaining and turned onto the final approach and with sixty degrees of bank attitude still existing he pressed the left rudder against its stop, moved the stick to the right, and side slipped to the right into the light cross wind towards the landing while reducing the throttle to the idle position. The unmistakable sensation of powerless flight was felt in all of the controls as the airplane seemed to have lost all life, was dying, was ready to give up flight. It was as though some power within the body of the pilot was lost. In over two thousand five hundred hours of flight this was Dave's fourth engine failure.

There is a "saying" among pilots that goes, "If you do something and you don't like the result, then un-do the last thing you did." In actual practice he had found that no one had to recite that idiom to him. Its message seemed to come naturally, instinctively.

Instinctively he opened the throttle and life returned to the engine and the airplane. The engine was running smoothly, but he couldn't land with the engine at that high power setting. He closed the throttle and it quit running again. That horrible, helpless, lifeless condition returned as he continued the sideslip towards the runway.

Several more times during the approach the sequence was repeated until he and the airplane were landed. Then by repeatedly opening and closing the throttle he taxied the airplane to its assigned parking spot on the ramp. He thought, "This must be what it was like to taxi a Sopwith Camel during World War I with its Clerget rotary engine that used ignition to control engine speed. The French LeRhone, Gnome, and the German Oberursel rotary engines used the same technique.

None of the spectators ever knew that he had suffered the engine failure on final approach during the show. He really didn't hear the applause from the crowd. An air show pilot seldom hears the applause. He was pre-occupied with concern for the condition of the airplane.

The following morning he removed the cowling from the engine and found nothing visibly wrong. A trial start-up proved that the fuel pressure was only reaching 4 P.S.I. and a Bendix P.S.-5 pressure carburetor requires a minimum of 9 P.S.I. to function properly. He was out of the air show for the weekend.

Feeling depressed he retrieved the toolbox from Sherry's car and began to remove the fuel pump and carburetor. During the approach and landing the preceding afternoon he had attempted to cover the fact that he was experiencing difficulties with the airplane and then, there he was publicly "wrenching" on the airplane.

"What's wrong, Dave? Can I help? What do you need?" Over and over, from nearly everyone that walked by. How could he proceed to make repairs while standing around talking to everyone about the problem? Finally the airport manager offered a hangar to store the broken bird in and it was pushed off the line and out of sight. Dave had no idea what a, "Pandora's box" was about to open for him.

On Monday Sherry took the fuel pump to the accessory shop and the verdict was, "Your fuel pump is, O.K."

Tuesday a long lunch period provided time to take the carburetor to a specialist who found a gray, sticky substance in the filter screen and poppet valve chamber. That long lunch period also earned a few sideward glances from the boss.

Wednesday Sherry retrieved the pump and left twenty- two dollars while Dave took another long lunch period and exchanged thirty- four dollars and ten cents for the carburetor. He was also questioned by his supervisor at the office, Tri Moga.

Wednesday evening he drove back to Howell and reassembled the airplane and found that he could only obtain six pounds per square inch of fuel pressure.

Thursday morning Sherry and he drove to the location of the airplane and devoted another day of effort into making a repair and still could only obtain six pounds per square inch of fuel pressure.

In despair he removed the carburetor and returned it to the shop and cancelled the air show commitments for the Thursday, Friday, Saturday, and Sunday.

Friday morning Dave and his son, Mike drove to the Howell airport where the airplane was hangared and removed the fuel tanks, fuel lines,

The Spirit's Journey

and other fuel system components. That effort required another full day of "wrenching".

Saturday he washed out the fuel tanks with Methyl Ethyl Ketone and Acetone followed by steaming them. This was done to remove the tank-sloshing compound that was initially put in to seal the rivets that retain the baffles in the main tank. Once the sloshing compound was washed out of the main tank a friend and welder, Ed Kilansky welded over the rivet heads to seal them. That cost two men eight hours each and all Saturday afternoon.

Sunday was spent reinstalling the fuel tank, fuel lines, gascolator, and fuel control valve. Then the header tank was found to leak, so he removed the header tank and on Monday morning left it with Ed Kilanski for the repair.

On Monday evening after working all day he retrieved the repaired header tank and drove to Howell and re-installed it. That meant he was back at home and in bed after midnight.

Finally, on Wednesday he caught a ride after work to Howell with another gentleman from the office who lived in the vicinity of Howell and flew the bird home to Grosse Ile.

The repairs seemed to be fine. He had no more difficulties with those systems, but other thoughts kept nagging at him. They were, how did the gray substance get into the fuel system?

Answer – Suspension in the fuel. The gray substance was fuel tank sloshing compound that he had to wash out of the tanks and lines. Somehow, the thought of toluene in the fuel had been raised.

Toluene was used in place of lead in the 100 octane aviation gasoline as a pre-ignition suppressant hence, the name 100-LL aviation gasoline that was brought onto the market by the oil companies in 1978. Toluene is also used as paint thinner, therefore it will dissolve the fuel tank sloshing compound?

Within the same week an antique Lockheed 12 taking off at a neighboring airport suffered a simultaneous failure of BOTH ENGINES. All persons aboard perished in the resulting accident. There were questions about the fuel in the airplane that were never answered. The only things shared by both engines are air and fuel.

Some of Dave's friends wondered aloud about toluene. What will it do in a certificated conventional airplane? How could it get into an airplane except through the fuel? Who checks for it and how? If aviation gasoline distributed in our area had toluene in it, then who mixed it in and why? Did the oil companies do it? Who should inspect for it?

Yes, the oil companies did it because aviation gasoline was easier and cheaper to produce with toluene in it instead of lead. The Environmental Protection Agency was also involved in the decision. At later dates the sloshing compound had to be rinsed out of tanks on Piper Cherokee and Comanche aircraft. The Marvel-Scheibler carburetors had to have the floats replaced because they had cork floats coated with varnish. The toluene dissolved the varnish and the cork floats became soaked in fuel and failed to function. There was also the danger of engine compartment fires associated with that problem. There were numerous engine failures because of the problem and the FAA was allowed to "skate" on it. Dave is not the only person in aviation that has a lot of serious questions about it.

The Great Lakes N-88-SK continued to run and fly just fine, but it is safe to say that, "It ain't always fun."

PART XLIII

DETROIT CITY AIRPORT SHOW

July 13, 14, and 15, 1979 was the weekend of the annual air show at Detroit City Airport and the Airport Manager, Lillian Snyder, a member of the Michigan Chapter of the 99's, and a member of the President's Advisory Committee on Aeronautics appointed by President Nixon, and Vice-President of the Airport Advisory commission had established what almost became a tradition at Detroit City Airport, the Annual Air Show, but with no accidents having occurred since its beginning in 1974 the TV stations and the newspapers in Detroit published very little notice of the event. The crowds of spectators were rather small for a city with a population of approximately one million and the airport being in the geographic center of it. There was adequate financial support from the City of Detroit and it featured the top performers in the United States and occasionally some from Canada.

Dave's parents, Jewel and Mac drove to Detroit from Alabama and brought his brother, Jimmy along. Their first objective was to see the house that Sherry and he had bought in Wyandotte and the second was to attend the air show.

The show was to open each day with demonstrations by the "Indian City Radio Controlled Model Club" with their model airplanes flying between the spectator line and the parked airplanes to be flown by the air show performers. They caused one of two moments of concern during the show when one of them almost struck Dave's Great Lakes and then came even closer to Jon Lynch's Citabria that he was going to use in the "Clown Act".

That moment of concern was more than adequately compensated for in Dave's opinion by the list of performers he was associated with. First, was the Detroit Police Department, Aviation Division doing a Fly-By and a rappelling demonstration from the helicopters. Second, came a fly by of the "warbirds" followed by a demonstration of the U.S. Army's Parachute Team, the "Golden Knights". Fourth, was the "American Flag Parachute Jump performed by the Golden Knights with Captain Bill Barber flying his Boeing B-75 and Bob Lyjak flying his Waco "Taperwing" circling the jumper trailing the flag on the way down. Fifth, was a solo performance by Duane Cole who was most often considered to be the, "Dean of American Sir Show Pilots". Sixth, was Airline Captain Jim Mynning performing his J-3 Cub Flying Safety Demonstration, a "dead stick" landing with some aerobatics prior to the landing. Then the seventh performer was Dave McKenzie presenting a solo aerobatics program in his Great Lakes biplane. Number eight was the "Car to Plane Transfer" performed by Bill Barber flying the J-3 Cub airplane with Jim Mynning driving the Corvette car and stuntman, Eddie Green doing the transfer. At the point in the run down the runway when Eddie Green, riding on the front fender of the Corvette grasped the rope ladder suspended from the airplane and was sliding from the car to the ladder nearly lost his handhold on the ladder it was a breathtaking moment for the performers that saw it, but they didn't know whether the spectators realized what had nearly occurred or not. Jon Lynch who played the part of the student pilot who inadvertently gets away with the Citabria presented the ninth act. It was the classic "Clown Act" and thanks to the announcer, Danny Clisham it pleased the spectators.

Oscar Boesch, a World War II veteran pilot of the German Luftwaffe presented an aerobatic performance in a sailplane accompanied by a recording of the song, "Born Free" for the tenth act that pleased both the spectators and the performers.

The tenth performance was a solo aerobatic act by Daniel Heligoin on Saturday and then his partner, Montaine Mallet flying the CAP-10, French built monoplanes on Sunday afternoon. Both of them were French citizens who had been trained and then emigrated from France to the United States. The eleventh act was again Captain Bill Barber flying his Boeing B-75 with Eddie Green walking the wing. Number twelve was Professor Bob Lyjak performing his very impressive exhibition of aerobatics in the Waco "Taperwing". The "Golden Knights" presenting a "Free Fall Baton Pass" followed Professor Lyjak's act.

The thirteenth and final demonstration was by the U.S. Marine Corps' presentation of a performance by their AV-8 Harrier, "Jump Jet".

On each of the two days the show ran three hours long without a hesitation or, "dead time" in it. Eleven of the participants were known internationally for their air show acts and accepted as, "the top rated talent". Dave felt that he was close to realizing one of his ambitions and was simultaneously humbled to have had the opportunity to appear with such an accomplished, recognized, and respected group of performers.

PART XLIV

OPEN HOUSE AT NEW HUDSON

Lou Spanberger, one of the Aerobatic Club members had bought the airport at New Hudson Michigan, mowed the grass, and cleaned it up. Lou decided to have an "Open House" and invite the residents of the community and anyone else that heard of the event. For entertainment he invited the Aerobatic Club and asked a few of the members to fly their aerobatic routines. Roger Fassnacht, John Gardner, and Dave McKenzie agreed to provide some entertainment on Saturday September 22, 1979.

Dave decided to add another figure to his air show demonstration, a "tailslide". He had been hesitant about performing them in public because of the results that were yielded by the first one he attempted on September 5, 1978.

He had decided to attempt a lomcevak, a dramatic tumbling maneuver begun very similarly to performing an outside snap roll on a forty-five degree climbing line. Then the pilot simply waits until the inverted spin develops and recovers from that. He was surprised at how gentle it seemed to be in the modified Great Lakes and how little altitude was lost in the maneuver's performance. "O.K.", he thought. "Let's see how the tailslide goes."

The lomcevak and the tailslide share the capability of applying forces to an airplane's structure that repeated exposure to can lead to subtle and slowly developing problems that are cumulative in nature and when the ultimate failure occurs it is likely to be disastrous. For that reason Dave had redesigned the ailerons, the vertical and horizontal stabilizers, the elevators, and the rudder, and selected the "tank" version of the W-670 Continental engine using a wooden propeller. He knew that he was going to be performing such maneuvers and wanted the airplane to definitely be capable of tolerating such abuses.

He selected four thousand feet AGL as an entry altitude, climbed above it, dove down to it, and pulled the airplane to the vertical attitude with the throttle set at "full power". Then he reduced the power to approximately fifty percent and waited for the airplane to come to a stop. Shortly before he thought it was going to stop moving he pushed the nose down ever so slightly and gripped the stick with both hands, braced his forearms by putting his elbows against his ribs, and braced his feet against the rudder pedals. He had no difficulty recognizing the backward movement of the airplane and the reverse flow of air across its surfaces, and then it abruptly pitched nose down and continued the swing until it was falling inverted which he easily recovered from.

The last time he had noticed the altimeter it was passing the reading of five thousand feet and was still moving upward. He hadn't really tried to conserve altitude when recovering from the tail slide, but he was astounded to see the altimeter indicating three thousand five hundred. He had fallen at least fifteen hundred feet from the apogee of the maneuver.

During his practice flights for the next year he regularly performed at least one tailslide and never again recovered at less altitude than the entry altitude. On September 22, 1979 he decided to add it to his repertoire.

Mike, Dave's son was the announcer for the show and everyone's performances went without any surprises to them and that included Dave's performance with the tailslide added to the list of maneuvers demonstrated. The show was successful and nothing was ever heard about it again.

On September 8 & 9, 1979 Dave and his Great Lakes were expected to return to Marion, Ohio for the annual Mid-Eastern Regional Fly-In. Just as it had been in '78 it was dealt with as a family venture and Mike flew down on Saturday morning with Dave while Terri rode down with Sherri in the Fairmont sedan. This time Dave was to fly the opening act of the show on both days. It went the way an air show is supposed to go, no problems, no disagreements with anyone, and no disappointments.

PART XLV

FALL CARNIVAL AT CLINTON

The Clinton Township Chamber of Commerce or Rotary Club, perhaps both were sponsoring its annual "Fall Festival" which consisted of a parade on both days, fire works on Saturday night, and a street carnival on Saturday afternoon. They wanted Dave to fly a low level aerobatic demonstration over the High School football field on the afternoon of September 29, 1979. They wanted to open the demonstration with two parachute jumps into the football field prior to Dave's flight, so some coordination with the sky diving operation from Meyers Field at Tecmseh, Michigan became essential.

Harold Lange had operated the sky diving school and parachute packing and inspection business at Tecumseh since the early sixties and was considered to be safe, reliable, and conscienscious. Dave had been assured that the skydivers understood they were to make their jumps at 4:30 prior to his performance. Hence, he planned to await the descent of the skydivers before starting.

Prior to the planned demonstration start time of 4:30 in the afternoon though, Dave was told to go ahead and start the performance. The skydivers would follow him. At 4:25 Dave took off from Drexel Scott's private runway on the west side of the community, climbed out and was prepared to begin the performance when he just felt uneasy.

He couldn't see the plane loaded with the skydivers and he had been told to disregard the original plan and start at 4:30, but he still felt uneasy. He started the performance and after completing the first couple of maneuvers he stopped the performance and looked above. There was the Cessna.

In a very short time two skydivers were out and trailing smoke from the canisters on their boots. Dave flew out to the east and circled until the skydivers were on the ground and then resumed his performance.

Dave, Mike, Sherry, and Terri spent Saturday night with Jan and Joe Rayne with the airplane in their hangar behind the house adjoining Drexel Scott's runway. When he left the next afternoon he took off on runway 18, turned left and happened to be flying east over U.S. Highway 12 that runs through downtown Clinton and passed over the parade that had just started. The club sponsoring the event thanked him for the little extra bit he had inadvertently provided at no cost.

He never heard anything from anyone about the confusion in the schedule, but he decided that the most dangerous thing about the air show business was the people who attempted to run them.

"Flying is many, many, many hours of pure, simple boredom with an occasional moment of stark terror and horror thrown in to revive the interest." is an idiom that he heard somewhere along the way.

PART XLVI

WINTER – '79 THROUGH "80

Sherry had taken the declaration of his intention to, "move out of this barn" that they owned in Wyandotte seriously. She participated in the search on Grosse Ile that had the potential to satisfy both of their objectives. Hers was to live in the "Down River" area of Detroit because it was close to the areas where the residences of her friends from Baptist Park Schools were located and she had acquired employment at Baptist Park Schools as a secretary in the Principal's Office. Grosse Ile met Dave's objectives because of its reputation as the "highest class" suburb located in the "Down River" Detroit area. Grosse Ile also had the airport that had been the, "Grosse Ile Naval Air Station" that was later called, "The Grosse Ile Naval Air Reserve Station". It was also the airport where he rented a "T" hangar and based the Great Lakes biplane. When the Fixed Base Operator, "Grosse Ile Flight Service" learned that he was a flight instructor he joined them and started flying part time at Grosse Ile Airport. It wasn't new. His second association with a flight school had been at Grosse Ile in July 1972 after he was called and hired by an organization named, "Arnold Aviation". Dave had only worked with Bob Arnold for six months in '72 and then severed any association with him. He was soon instructing at two airports, Detroit Metropolitan and Grosse Ile.

In November 1979 they found a ranch style home with an attached two-car garage and a finished basement at 21141 H.C.L. Jackson Drive, purchased it, sold "the barn" located in Wyandotte and on December

15, 1979 they moved to Grosse Ile. Dave thought that he'd spend the rest of his life at that address.

Of course it was just as difficult to move the second time as it was the first time, but Dave liked Grosse Ile, based the Great Lakes at that airport, and was employed part time as a flight instructor on the airport. Sherry and Dave had everything organized, reasonably convenient, and comfortable. Life was good.

Dave's students at Detroit Metropolitan Airport were Tony Malizia, Gary Clark, and Dr. James Fordyce who were all flying with the Southgate Flying Club. Other students were Mike Gobb and Ilene Hemingway flying from Grosse Ile Airport. Kathy Holliday, wife of Dick Holliday who was one of the owners of Grosse Ile Flight Service was also one of his students. Dick Holliday had been a Marine Corps jet fighter pilot and was instructing at Grosse Ile and asked Dave to be his wife's instructor. The lesson that Dave learned from Dick was, no father should teach a son, mother teach a daughter, sister teach a brother, or husband attempt to teach a wife anything. It won't work. Dick knew that.

Mike Gobb was a basic student, the son of another employee of Ford who knew Dave at Ford Motor Co.. Mike soloed under Dave's tutelage and had almost completed his training when it was interrupted by totally unexpected events. He did complete his training, earned his Private license, graduated from Western Michigan University, and became the manager of the Alpena, Michigan Airport.

Ilene Hemingway earned her "tail wheel" endorsement, her instrument pilot rating, and her Commercial Pilot Certificate taking instruction from Dave that winter and spring.

The last flight of the Great Lakes in 1979 occurred on November 18 and was just a pleasure flight that is logged as, "acro practice". The first flight in 1980 was on January 20 and was in formation with a Bellanca 8KCAB flown by Kathy Harris to take photos of the Great Lakes in flight. The next flight of the airplane was on April 6 after completion of its annual condition inspection followed by three flights on April 13. It was flown again on April 20 and on April 27 a crack developed in the oil tank. On May 3 Dave gave friends rides in the airplane, and practiced aerobatics on the 4th and 5th. On the 5th the engine consumed or blew 6 quarts of oil in fifty minutes of flight.

Dave's response to that development was to perform a top overhaul on the engine and the next time it flew was on the test flight on June 9 followed by some aerobatic practice on June 12 and 13.

June 14 he and Mike flew it to Owosso for the Michigan E.A.A. State Convention and air show where Dave performed on the 14th and 15th returning to Grosse Ile on the 15th.

On June 19 at 6:00 o'clock in the morning he met two "Playboy" magazine centerfold bunnies for a photo session at the south end of Grosse Ile Airport for a photo session with a hot rod named, "The Little Red Baron" that was built for appearances in car shows. The top of the car simulated a World War - I vintage German infantry helmet and there were two chrome plated machine guns on the hood. The girls performed their "changes" in a motor home that accompanied the hot rod and the trailer that it was transported in.

Of course Sherry, Terri, and Mike attended the photo session and Dave took each of the girls for a ride around the island following the photo session. Dave did learn which issues of "Playboy" magazine the girls had been featured in and obtained copies. While he was hospitalized at a later date the copies of the magazines disappeared. They were the only copies of "Playboy" that he bought in his entire life.

On the weekend of July 11, 12, and 13 Dave was very busy acting as the, "Contest Director" for the IAC-88 Annual Aerobatic Contest in Owosso, Michigan, the fourth year in succession that he had done so and he had a simultaneous commitment to fly an aerobatic exhibition at a "Dawn Patrol" at Jewett Airport in Mason, Michigan. His intentions were to assist the contest officials in starting the contest in Owosso early in the morning of the 13th and after the event was in operation he would fly with his son Mike to Mason, fly the performance with Mike announcing, and then return to Owosso and complete his duties at the aerobatic contest. It was really a very ambitious weekend.

Perhaps it was overly ambitious and Dave may have been assuming more responsibility than he should have assumed.

PART XLVII

LOST AND DISORIENTED

The memories were very sketchy, sporadic, and vague. Several days after July 13, 1980 Dave awakened in a darkened room, alone, and in a silence unbelievably deep. He realized that he was not aware of any sensitivity in his legs and he couldn't raise his head to visually locate his feet. He wondered where he was and decided that it had to be in a hospital. Then he became aware of the pain in his right leg and soon knew that something was on and in his nose. He decided that nothing really mattered. He'd take care of it later. He felt fatigued and soon sleep overcame him again.

Someone was in the room when he awoke and he decided the other person must be a physician that was caring for him, but he neither knew nor cared. He went back to sleep. Sleep was peaceful, comfortable, safe, and secure. Nothing seemed to matter and he knew that he couldn't do anything, so he thought, "Just sleep. Everything will still be here when I awake."

As time progressed his waking moments became more frequent and each one was of a slightly longer duration. His curiosity finally revived and he was certain that he was in a hospital and he was trying to determine the cause of his hospitalization. He hadn't seen any demons, hadn't heard any music, beautiful or otherwise, and he was at peace with no concerns. He knew that he was incapacitated and would be unable to affect the outcome of anything for sometime, if ever. He resigned himself to accepting the situation and started making attempts to figure out what had occurred that caused him to be in this situation. He considered and eliminated an automobile accident, some physical problem like a stroke, a disaster of some type, and decided to review his lifestyle and try to find something that could have caused this. Finally it occurred to him that he must have suffered an airplane accident. He thought that if it were an airplane accident then the Great Lakes would be gone. He deliberately chose to force himself to mentally accept the loss and to "move on".

He finally recognized Sherry when she appeared on one of her visits and she confirmed that he did have an accident in the Great Lakes. He was also aware that Sherry was attempting to feed him like a baby and kept dribbling the food on his chin and then attempting to wipe it off with the spoon. He found all of that annoying, but could do nothing about it.

Later he learned that he had numerous other visitors like his mother and father, friends from the aerobatic club, and from Ford Motor Company. One visit was from a Pastor Jack Downs from the Gilead Baptist Church and Roger Cook, the Principal from Baptist Park Schools where Sherry was employed and the children attended, but his memory didn't record their visits.

Psychiatrists and Psychologists later told him that there are unproven theories about the memory indicating that intolerable trauma causes an area of the brain to store memories until the sufferer can withstand the stress that would be caused by the recollection of the trauma and the memory lapse begins at some time prior to the date of the trauma and does not release information gathered for some time after receipt of the trauma The theory proposes that with the passage of time the impact of the trauma becomes more tolerable and the memory bank will release the information to recollection from some time prior to and an equal time after the receipt of the trauma, but not the causal event itself until the victim is physically and mentally capable of enduring it. It may never be remembered.

Dave later learned that he was unconscious for at least a week, possibly two weeks. The medical professionals refused to reveal the time to him. Then he began to slowly regain the ability to speak and to recognize other people. Sherry, Terri, and Mike were invited into the Psychiatrist's office at the hospital and advised that the CATSCAN taken three days after the accident indicated that there was some bruising and edema of the brain that was not unusual in severe impacts, but such patients recovered over the passage of time. He went on to state that they were concerned in Dave's case because he spoke with a slur that resembled a southern accent and were concerned that it was a symptom of brain damage.

The staff would sometimes enter the room and ask, "Dave, where do you live?"

His response was, "One quarter mile north of Michigan Avenue and about three hundred feet west of bitch daily". He was speaking of the home location at 26034 Woodbine Drive in Inkster located west of Beech Daly that he had moved out of in July of '78. He did not at that time recall the move to Wyandotte or Grosse Ile. There was a young nurse on the staff that he must have enjoyed teasing. When she was on duty and he wished for her assistance he'd ask, "Tell that waitress that I need - - - -."

The nurse would enter the room saying, "David, I'm NOT A WAITRESS. I didn't spend all that money and time going to school to be a waitress. I'M A NURSE."

He was in a cast extending from the pelvic area to the throat because he had suffered compressive fractures of the vertebra T-8, T-12, and L-1, fractured two ribs in the right side and the right femur. Herrington rods had been implanted in his back and a stainless steel rod connecting a cast around the upper arm to another cast around the forearm with the elbow exposed restricted movement of his right arm. Its purpose was to eliminate bending of the arm until a severe laceration on the elbow healed.

At the time of impact with the ground some of the cockpit coaming and the windshield had struck his forehead in the edge of the hairline and torn his scalp lose across the front and backwards along both sides of the skull and the scalp had flipped backwards like a lid. The plastic surgeon had made that repair on Sunday evening July 13.

Three weeks after the initial surgery was performed a titanium rod was implanted in his right femur and retained with three screws to repair its fracture. The staff at Ingham County Medical Center had decided to postpone that surgery until he had partially recovered from the other repairs due to concern about stress from performing so much surgery in one short period of time. The fractured femur had been the cause of the discomfort in the right leg that Dave experienced since he first regained consciousness.

Approximately September 1 the body cast was removed and a fiberglass body brace was made for use when Dave would later attempt to reach a sitting position. The prognosis was that he would never walk, stand, serve an employer, or perform any useful function again.

His son, Mike had ridden from the accident site to the hospital with the Ingham County Deputy Sheriff who escorted the ambulance. Sherry and Terri were flown to Capitol City Airport at Lansing by the Lacy family in their Cessna 310 and then driven to the hospital by the Ingham County Sheriff's Department. Jewel, Mac, and Jimmy were en route from Alabama to Grosse Ile for their first visit with Sherry, Dave, and the grandchildren since their move to Grosse Ile and were expected to arrive on the afternoon of July 14. News of the accident had spread rapidly around the aerobatic contest site, Owosso and had been relayed to the Grosse Ile Flight Service office where friends began to think of ways to be of assistance to the family. Kathy Harris and Fred Ahles went to the McKenzie home on July 14 and remained with Terri and Mike until the arrival of Jewel, Mac, and Jimmy on the 15th and Mike's informing his grandparents of the accident that occurred on July 13, Jewel's birthday, the same date as her son's accident.

On Friday, July 18 Mike left Grosse Ile with Jewel, Mac, and Jimmy for Evergreen where he spent the next few weeks. The Rayne's, Jan and Joe who were childless took Terri into their home and Jan liked Terri very much anyway. Terri and Jan both loved horses and Jan had three.

During their visit to the hospital where Dave was lying unconscious on July 15 or 16 Gilead Baptist Church's Pastor, Jack Downs and the Principal of Baptist Park Schools, Roger Cook advised Sherry that they would recommend that she submit her resignation of the position she held in the Principal's office in order to devote her attention to her husband. That led her to apply for admission to the Grosse Ile Public Schools for Terri and Mike that was accepted, of course. When Dave learned of the change he was elated and considered it to be the only good thing to appear as a result of the accident.

The Spirit's Journey

On September 15 a twin-engine air ambulance flown by Jeff Bush and Dave Ahles with Larry Batha as the medical attendant flew Dave from Capitol City Airport to Grosse Ile Airport and he was transported by ambulance to Wyandotte General Hospital for recovery and rehabilitation. Grosse Ile Flight Service where Dave was employed as a part time flight instructor provided the air ambulance and the crew flew it voluntarily.

Dave had never discussed the possibility of such events occurring with his friends, fellow employees at Ford and the airport, or acquaintances, and popularity was not a concern of his. He had always accepted the fact that most people are courteous, considerate, thoughtful, and are helpful to their friends, acquaintances, and others when stress enters their lives. In later years he realized that he had witnessed the truth in those thoughts, felt justified in his opinion, and was sincerely thankful even though he didn't believe there was any way that he would ever have the ability to tell them of his appreciation.

During the admission to Wyandotte General Hospital, Rehabilitation and Physical Medicine Department he was asked for his home address and gave them the address that was out of date by three years, was admitted, and placed in a "ward" with four other patients because it was located immediately across the hall from the nurse's station. His weight at admission was 109 pounds. He weighed 150 pounds prior to the injuries.

Dave did not like the ward at all. He had been placed there because of its proximity to the nurses' station and he knew that it was for their convenience. He had no television, no telephone, no privacy, and was constantly being disturbed by the other patient's visitors and needs. It was not a place to find peace, rest, and recuperation in Dave's opinion, but it had been touted as one of the best rehabilitation facilities in the Detroit area and on a par with The University of Michigan Hospital and was only three and a half miles from the home on Grosse Ile.

After three weeks in "Rehab" he was taken to the Physical Therapy Gym and met Brenda Laravee, a therapist who admitted to being, an "Air Force Brat" who started him out on a tilt table, a table that the patient was strapped onto and then rotated towards the up-right or vertical position. The intent was to strengthen the heart muscles that had atrophied from merely circulating the blood stream in the horizontal plane and didn't need to push it from the feet to the head as in standing. Dave's heart had atrophied in the two and a half months since the accident.

Miss Laravee began the program by rotating Dave from the supine position to an angle of about thirty degrees from the level position. The heart rate would go up to 130 beats per minute where she'd keep him for 60 seconds, and then rotate him back to level, let the heart rate return to about 80, and do it again. That was the beginning of the physical therapy that lasted one hour per day for about two weeks.

At an examination by one of the physicians Dave grew rather demanding and wanted to know why therapy was only being done one hour per day. Why not twice per day? The schedule was increased.

During the second week of October he was removed from the bed after lunch one day, literally dropped into a wheel chair with the back reclined to an angle approximately thirty degrees from the horizontal and taken to a room in the adjacent building where the technician approached him with something that appeared to be a pizza cutter and sawed a hole in his scalp, used a piece of "goo" to attach a wire to the wound, and then repeated the procedure thirteen more times. He then told Dave to relax and to avoid thinking about ANYTHING.

Dave went to sleep and was awakened by the sensations caused by the glued in place wires being pulled free.

The next day the entire sequence of events was repeated. The procedure was called, "an EKG" and is supposed to give an indication of the brain's ability to conduct electrical currents from one area to another. The current flow is recorded on a graph by an ink stylus that is mounted in a recorder that would remind an observer of a lie detector.

About two weeks later Sherry, Terri, and Mike were called into the Psychiatrist, Dr. Amburg's office and told of the test that disclosed no abnormalities. Dr. "Ari" Amburg then added, "But have you noticed how Mr. McKenzie speaks with a sort of a slur, sort of like a southern accent? I suspect some brain damage."

Physical therapy continued directed by the very competent Miss Laravee and by that time Dave had been moved to a semi-private room. He had television and a telephone and every evening Sherry, Terri, and Mike would arrive for a visit from 6:30 until 8:00 o'clock and bring the day's mail and any book he had requested out of his collection.

The Spirit's Journey

The lady who had been his, "Occupational Therapist" had asked what his profession had been. When she was told that he designed automobile chassis it was obvious to Dave that she had no idea what he was talking about, so he volunteered to contact his supervisor at Ford Motor Co., "Tri" Moga and ask if he would give her a tour of the department and give her an opportunity to view the job's physical capability requirments. Mr. Moga agreed to do just that and Dave advised the Therapist of his concurrence with the plan.

She said that she presented the idea in a staff meeting and it was forwarded to the Director of Rehabilitation and Physical Medicine, Dr. Joe Guyon. There was never a response to the idea offered.

On the first Thursday in November two nurse's appeared in the room after lunch, pulled him out of bed, and dropped him into the wheel chair and rolled the wheel chair down the hallway to the room referred to as the, "day room". Soon all twenty-six patients in the "Rehab Wing" were parked in their wheel chairs around the perimeter of the room listening to Dr. Joe Guyon introduce Dr. "Ari" Amburg. (His given name was, Erik, but he personally preferred, "Ari".)

After the introduction Dr. Amburg asked each of the patients to, "introduce themselves, explain why they were here, and state their goal." The first patient asked to respond to the request was on Dave's left side, so each patient attempted to follow the instructions and the exercise became extremely embarrassing. At least Dave thought most of them were embarrassed like a lady named, Anna who said, "My name is Anna. I don't know what happened. I want to pet my doggie." It was obvious that she was seriously challenged to put the three thoughts together and to annunciate them. Dave also thought that it was unprofessional, cruel, and inconsiderate for the two physicians to put that much stress and embarrassment on patients who were mostly victims of strokes. Some of them had suffered heart attacks, and one was the victim of an industrial accident and like Dave had a broken back.

Dave responded when his turn came with, "I'm Dave McKenzie. I'm here because I flew an airplane into the ground and I have three goals. One is to get out of here. The second is to get back to work, and the third is to get back to flying."

Someone gripped the handles on his wheel chair, turned it to the left, and pushed him back to his room where Dave was assisted out of the chair and back into the bed. Dr. "Ari" Amburg appeared for a visit at the end of the hour and Dave attempted to explain what his impression of the opening of the session was. The tone of the conversation degenerated from that point on and Dr. Amburg left the room as Dave was talking.

Later that day Dr. Joe Guyon visited Dave for a very short visit during which he said, "You should give up any idea of flying or working again. Some day you will realize that you cannot physically do such things and when you do it will be disastrous. The sooner you accept the idea that you can't, the better for it you will be. You will never again see an assembly line as long as you live." He turned and left the room before Dave had any opportunity to respond. Dave thought that the man obviously allowed his own ego to limit his capacity to positively affect the outcome of the treatment of patients at the facility. The man just couldn't accept the thought that Dave was a designer.

Dave was being given therapy twice per day and was slowly improving, but that improvement was torturous and slow. On December 24, 1980 he was discharged and told that in January he would become an, "outpatient".

His mother, father, and brother drove from south Alabama to visit the family over the Christmas and New Years holidays and even though there wasn't a very good mood in the home because Dave was having so much difficulty getting around. It was still an improvement over the environment in the hospital.

In January 1981 Dave was scheduled with Brenda Laravee for one-hour sessions in the Physical Therapy Gymnasium from 10:30 to 11:30 on Tuesday, Thursday, and Friday mornings. She advised him that if he would come in at 10:00 she would have him attempt standing "in the bars" for a half hour before logging him in for the session. In other words she was going to give him thirty minutes per session at no charge.

She had him attempting to stand between the parallel bars and remain standing after removing his hands from the bars. The objective that she assigned was for him to stand for five seconds with his hands above the bars and without touching them. At that time he could stand for approximately three seconds before beginning to topple over. She also assigned Dave several exercises to perform on the exercise mat that

The Spirit's Journey

was on a table approximately eighteen inches high, eight feet long, and six feet wide that required her assistance. Dr. Mary Ann Guyon was the Supervising Physician and prescribed those exercises. (Their home was the fifth house south of the McKenzie's on Grosse Ile.)

On every Tuesday, Thursday, and Friday morning Sherry and Dave would leave their home at 9:30, Sherry would drive to the hospital where Dave would get out of the car and using crutches would enter and cross the lobby enroute to the gymnasium where he'd remove his coat and proceed to the parallel bars and begin his attempts to stand and maintain his balance. After a half hour of that exercise Brenda would have him move to one of the mats, give him the appropriate exercise device and then assist or guide him in the exercises.

Sherry would drive to the Grosse Ile Post Office, retrieve the mail for the Grosse Ile Baptist Church and the McKenzie's, stop at the church, and then return to the hospital at 11:30 to meet Dave and drive him home. One morning as Dave was crossing the lobby enroute to the gymnasium Dr. Mary Ann Guyon stopped him, stood in front of him, and began to interrogate him using a very strong tone of voice.

"Why don't you get a wheel chair? You're using too much energy attempting to walk. You're going to have a heart attack." Dave didn't believe he could provide an answer that Dr. Guyon would accept. She had obviously decided that there was no hope for any level of recovery and she was not alone in supporting that diagnosis.

Dave was angered by the conceit and indifference to a patient's welfare that he detected in the operation of the facility. He thought, "These people don't assist patients in recovery. They encourage them to cease attempting to return to useful lives. Why? Was this procedure an alternative to a patient failing after having been encouraged? Was it a defense to a possible charge of misdiagnosis?" Dave never discovered answers to those questions.

Brenda Laravee was called into the office of Dr. Mary Ann Guyon and asked, "Why are you wasting time having that patient stand in the parallel bars? He is never going to stand and you are wasting time and effort uselessly." Brenda wasn't intimidated and the exercise program continued in the manner that she wanted to run it. There was one free thinker in the facility.

Strength and balance improved and Dave was moved from the full crutches to the Loftstrand crutches. His right leg was not strengthening, so a brace was fitted to it that fastened around the thigh with a Velcro strap and another one around the lower leg. Rods were located on the outside and inside of the leg and each had a hinge at the knee that would lock in the straight position with a bar behind the knee to be used as a trigger to unlock the brace and allow the knee to bend. With all of that hardware Dave was able to move around reasonably well and had started using stair steps.

He was feeling so confident that he telephoned "Tri" Moga and asked him to visit the personnel department, obtain the forms for a physician's "Return to Work" order and forward them to him. Mr. Moga seemed pleased to perform that favor and Dave left them with Dr. Joe Guyon's receptionist.

XLVIII

ANOTHER OBSTACLE

Jewel, and Mac drove up from Alabama on Saturday, the 6th of June 1981 to visit Sherry and Dave. There was no way that the visit could be compared to the visit during the prior holiday season. The situation had improved considerably. Sherry was again earning wages but, as an employee of the Grosse Ile Baptist Church, not Baptist Park Schools. Terri and Mike were students in the Grosse Ile Public Schools, the third ranking, scholastically in the State of Michigan, and Dave was talking about going back to work.

On the 4th of June he had submitted to surgery on the muscles in his right eye to correct the esophoria and wore dark glasses any time that he went out of the house to protect it from the glare. Otherwise he was doing well and looking forward to receiving approval from Dr. Joe Guyon to return to the office and the resumption of his duties. Of course he would have to continue receiving physical therapy and on the morning of June 11th Mac intended to drive him to the hospital while Sherry and Jewel went their own way.

Dave opened the passenger's door of Mac's fordor Ford "Galaxy" sedan and couldn't find a seat belt. After feeling along the opening between the seat and back cushions he finally found the belt, pulled the filthy thing from its hiding place, stretched it out on the seat, got in the car and fastened the dirty belt around his pelvis. He had been very religious about that habit and had installed safety belts in every car he had owned prior to the mandated installation of them in the early sixties. He was a firm believer in their use.

Mac was already in the driver's seat and had started the engine. They drove north to Bridge Road and west across the toll bridge then north on Biddle Avenue. During the ride Dave kept his head down to avoid the bright light of day because it was painful after the eye surgery that he had endured the previous Thursday. He would look up occasionally to monitor their progress and offer directions to Mac. He was looking down when he heard the sound of an air horn. He looked up and instantly saw that they were in the center of three lanes northbound on Biddle and were surrounded by other traffic. He looked to the right and recognized the abandoned Firestone Tire plant and saw the locomotive on the tracks at the north end of the plant exiting the property and entering the crossing of Biddle Avenue.

His first reaction was to say, "Stop, stop!" and as soon as he heard himself say it he was wishing that he hadn't made a sound. Mac released the pressure on the accelerator pedal and Dave was thinking, "If I hadn't demanded a stop causing dad to release the accelerator pedal we might have made it across, but we're not going to now." Actually, they probably would not have cleared the locomotive.

Dave watched the impact of the train's "cow catcher" against the car's right front fender and saw the rear of the hood beginning to fold upwards and forwards, as the windshield became thousands of small glass pebbles. He looked to the right as the car slid across the front of the locomotive and as the "cow catcher" was smashing the right side of the passenger compartment in he was making a concerted effort to stay off the door. The front of the car cleared the "cow catcher" as the train continued to push the rear of the car along the tracks rotating the car ninety degrees to its right. Very quickly the wreckage of the car became entangled in the running gear of the train and it dragged the vehicle along with it. By this time there were so many glass fragments flying around in the car that Dave decided he'd better close his eyes. Finally, there was silence except the sound of huge diesel engines idling. Dave thought, "It's a time like this when your sense of timing is confused and you had better keep your eyes closed a little longer."

He opened his eyes and the first thing that he noticed was the unusual angle formed by the junction of his right thigh and lower leg.

He knew there had to be a fracture in the knee and decided to simply await the next development. Mac was attempting to turn the ignition off and get the key out of the lock. There seemed to be a few scratches on his forehead and he asked Dave, "Are you hurt?"

Dave's response was, "Yes, I think I have a problem with the right leg." and he could see the guard for the gate into the "Wyandotte Chemical Co." property approaching the car.

The guard passed a handful of paper towels through Mac's window to Dave and said, "Don't move. Don't try to get out. The ambulance is on the way. Dave knew from the amount of blood running into his eyes that he had several cuts from the glass particles on his forehead and responded to the guard, "Don't worry. I'm not going anywhere."

The ambulance arrived and the attendants and Dave together decided how to transfer from the car seat onto the stretcher lying in the street. The difficulty in the transfer was caused by the right side of the car still being entangled with the train's running gear. After the transfer was completed Dave was asked which hospital he wished to be transported to and he answered, "Wyandotte, it's the closest and that's where we were going anyhow."

The ambulance driver replied, "No, we can only take you to Riverview. We're from the Riverview Fire Department. If you insist on going to Wyandotte you'll have to call another ambulance."

"Call the other ambulance." responded Dave.

The second ambulance that was called to the scene carried him to Wyandotte General Hospital and upon entering he asked the girl in the admitting office for the emergency room to call the Physical Therapy Gymnasium and advise Brenda Laravee that he'd be late for the appointment.

He was moved to a room and while waiting for a bed to become available in the hospital he was given "IV's" for sedation, pain, and fighting infections. During the afternoon an investigator for the Wausau Insurance Company, the railroad's insurer visited the room, took a verbal report, copied it down, and then asked Dave to endorse it. Because his glasses had been broken in the accident and the IV that sedated him Dave was unable to read it. The Insurance Investigator read it to him and then asked him to endorse it as being correct. Dave endorsed the document and the Insurance Investigator departed.

1973 Ford Galaxie after collision with locomotive (author's collection)

Dave awoke in the hospital the next morning, ate breakfast, and then made three telephone calls. The first one was to Sherry and he spoke to his mother, his father, and his brother. The second call was to his office at Ford Motor Co. His Supervisor, Tri Moga was not available, so he advised Jack Perrin, one of the other designers about the preceding day's developments. The third call was to one of his former students, Tony Malizia, an attorney.

He spent the next month in the hospital. The Drs. Guyon recommended to the Orthopedic Surgeon that he fuse the knee. The Orthopedist, Dr. Peter Palmer, after advising Dave that he was not working on pipes and lumber this time agreed with Dave. He could always wear the brace if the leg failed to regain muscle strength, but if the knee were fused the door was closed to most uses for the leg. Dr. Palmer repaired the Tibial Plateau fracture as best he could and after removal of some of the meniscus in the knee there was approximately an eleven degree bend outward of the lower leg at the knee. Dave could not place his heels together. So what!

He was dismissed from Wyandotte General Hospital on Saturday, July 4 and on Sunday Sherry drove the Fairmont with Dave sitting crosswise in the rear seat to Owosso for the annual IAC-88 aerobatic contest. Dave was very flattered when they stopped the contest for all of the contestants to welcome him to the event, but the highlight of the day came when Fred Leidig from Medina, Ohio finished his flight in the Intermediate Category and taxied his Hyperbipe biplane up to the lounge that Dave was laying in, stopped the engine, got out, walked over to Dave and said, "Look it over." He knew that Dave wanted a Sorrel SNS-7, "Hyperbipe".

PART XLIX

RETURN TO FORD

In one of several conversations with the attorney, Anthony "Tony" Malizia Dave learned that Tony had discussed the situation with one of the Drs. Guyon and wasn't pleased about their prognosis, so he had telephoned the Physical Medicine and Rehabilitation Department at the University of Michigan and spoken to Jo Gunn, the office manager.

She recommended that Dave make an appointment with Dr. Ted Cole, the head of the department and to bring copies of the "Physician's Return to Work" form for Ford Motor Company. It was in August 1981 that Dave was examined by Dr. Cole who said, "Yes, I'll write the order for you to return to work on October 1. It will specify that you will work only half a day at most for the first month, six hours per day in the second month, and return to a full eight-hour workday in December. If that works out O.K. then I'll remove all restrictions in January '82."

Dave was overjoyed and felt much better about the train accident. Had it not been for that incident he would have probably still been "stooging" around with Dr. Guyon attempting to obtain permission to return to the office and his duties.

He always felt very strongly that the Drs. Guyon were being overly cautious and pessimistic about his goals and ambitions. He felt that perhaps their hesitancy about offering any encouragement and cooperation might have been driven by a concern about the possibility of legal action and pessimism about the potential for success. After all, they had refused to accept the tour of the office and the job description that was offered by his supervisor.

On October 1, 1981 he returned to his place of employment using full crutches and enjoying use of an assigned parking spot near the entrance to the office. His starting time was set at 12:30 P.M., which allowed him to continue the outpatient physical therapy program with Brenda Laravee, and he thought, "Somebody up there likes me."

He had been out of the office for fourteen and a half months. A normal maximum absence for sickness or injury would be twelve months, but the Ford Motor Company management team had been extremely tolerant of Dave's situation. He really felt indebted to a company that would stand by an employee like that one had done. Dave felt that their holding the employment open should justify allegiance to any organization and the decision he had made that day in 1943 while in the first grade was one of the best that he ever made.

Physical Therapy continued without any surprises and gradually his job assignments became more complex and he felt that his skills as a designer were rapidly returning. He was also driving his station wagon without any special aids required to compensate for his handicaps and he was sure that he would return to the activities that he so dearly loved, flying.

In January he returned to a full time schedule without restrictions and had become accustomed to the limitations in access and activity dictated by his inability to walk without the assistance of crutches. That soon improved when he transitioned to using Loftstrand crutches.

In March he met with his supervisor, Tri Moga and asked for training on the computer graphics system that was coming into use in the company. The reasons he offered for requesting the training were drafting with lead holders, triangles, and compasses was being replaced by computer graphics and every time a designer retired a draftsman did not replace him. A computer station and operator replaced him. Dave pointed out that he was forty-five years old and only had seventeen years experience with Ford Motor Co. He had a long way to go and if he had any hope of remaining there he had better, "fit into the system". Mr. Moga assigned him to a class providing forty hours of computer training.

Upon completion of the computer graphics course he was transferred to a different department and told to begin design studies of a proposal to install a BMW diesel engine in the Lincoln Town Car. The job utilized the old tools and skills. It did not utilize the coming computer tools. Of course, the Lincoln car and the diesel engine were both designed in the old system and were not yet converted to computer graphics. Dave was pessimistic about the marketability of the concept because he just couldn't picture a person with the intelligence, talent, earning capacity, and social sophistication that matches those characteristics being anxious to take his wife and family into a truck stop to refuel his car while on vacation.

On April 1, 1983 Dave was one of thirteen designers assigned to a new department named, "Advanced Chassis" that was to start with a clean sheet of paper and design a replacement for the Thunderbird that was then in production. He was elated.

Dave enjoyed the slightly more than two years that he was associated with the designers, engineers, and management in "Advanced Chassis" more than any of the assignments that he had in the thirty nine years and eight months that he was an employee of Ford Motor Company.

A person suffering physical disabilities is likely to be the victim of accidents if they continue to participate in many of the activities he/she enjoyed prior to becoming disabled. Dave knew that, but was determined not to succumb to acceptance of the physical condition and continued to resume his life's course. On September 18, 1983 he dropped a piece of angle iron and fractured the first metatarsal in the right foot while working on the floor pan for another car project. He lost no time from Ford Motor Co. due to that injury, but on September 18, 1984 he took a fall while removing some parts from a junked Volkswagen to use on the same project. The fall fractured the right femur in two places, the left radius, and the left ulna. Those injuries required a medical leave of absence totaling three months.

In August of '85 the company decided to proceed towards production of the car for the '89 model year and six designers from "Advanced Chassis" were transferred with the project to the pre-production design activity. Dave was one of the six designers that moved with the project. He didn't work on the T-Bird again until he produced drawings of the installation of dual exhaust gas catalytic converters in the car for presentation to the California Air Resources Board in 1995. The drawings showed that four converters would not fit in the car. That was one of the reasons Ford Motor Company ended production of the T-Bird at the end of the '97 model year.

Of course nothing good exists without its disappointments and on the '89 'Bird project one huge disappointment was the loss of the Executive Engineer who had been directed by Upper Management to provide a design for the car utilizing rear wheel drive and featuring independent rear wheel suspension.

Every time a designer in the department wanted to put a lightening hole in a part or to use a lighter gage material with stiffening ridges the lighter part proposal would be rejected and the finished car weighed 4100 pounds, approximately 500 pounds in excess of the targeted weight. A "rule of thumb" estimate of the cost of an automobile at the time was one dollar per pound. That was the cost of the car without paying for the cost of facilities for the manufacture, engineering and design costs, purchasing and marketing costs, fringe benefits and medical costs for the employees, and many other overhead costs for operation of an automobile producing organization. Responsibility for that five hundred dollar cost increase above the target dictated by the refusal to utilize weight saving production and design techniques was assigned to the Executive Engineer in charge of the project and he was released from the company. Some employees should have seen it coming. American automobile design was about to become dictated by "bean counters" instead of being the result of designers and engineers applying mathematical principles to the laws of physics to produce a desirable and economic vehicle.

PART L

DESIRE OVERCOMES OBJECTIONS

In a meeting in the Day Room at Wyandotte General Hospital in November, 1980 Dave had said, "I have three objectives. One is to get out of the hospital. Two is to return to work and three is to fly again."

On Sunday afternoon, July 5, 1982 Sherry and he had gone to Dearborn to visit a Tile store and were considering remodeling the kitchen in the house and after completing the visit at the store decided to visit Mettetal airport since they were so close to it. They encountered Kim Kovach, Mary Lou Leverance, Al Wells, and "Doc" Kovach when they arrived at the airport.

Kim and Dave climbed into a Cessna 152 and flew it for a half hour. Then they boarded "Doc" Kovach's Bellanca Decathlon and flew some aerobatics for another half hour. It was the fourth time in his career that Dave had flown a Decathlon from the front seat and he was surprised by his ability to still perform the loop, the slow roll, the barrel roll, the snap roll, and the half reverse Cuban eight. He was really pleased and felt like he thought he would have after a two-week lay off from aerobatics.

In 1973 at the Experimental Aircraft Association Annual Fly-in and Convention at Oshkosh, Wisconsin he had seen the prototype Sorrel SNS-7 biplane parked in the line of Wittman Tailwinds that were on display and his first thought was, "Who in his right mind would attempt to make a biplane out of a Tailwind?" The card on the biplane identified it as the, "Hyperbipe" and except for it appearing to be a modified "Tailwind" Dave was impressed and liked it. That afternoon he saw it flying in the air show and decided, "That's not a modified, "Tailwind". That's a new airplane."

The next morning he went back to the Hyperbipe's parking place and inspected it closely. As he walked away from it he was thinking, "If my Great Lakes weren't fifty percent complete I'd build one of these."

When he was released from the Hospital in December, 1980 he waited until Jewel, Mac, and Jimmy had returned to Alabama and then had Mike bring his copies of the 1973-'74 issues of "Sport Aviation" magazine up from the basement, looked up the Sorrel's telephone number and called them. They had ceased to provide drawings, builder manuals, and material packages for the airplane, but assured him that they remembered his Great Lakes from Oshkosh in '78 and would support him if he were able to purchase an unfinished Hyperbipe project from one of their former purchasers.

On the Halloween weekend in '82 Sherry and he flew via Republic Airlines to Minneapolis, Minnesota, inspected a project that had been advertised as being for sale, purchased it, and brought it back to Grosse Ile, Michigan in a rented truck.

LI

LIFE CHANGES AGAIN

Sherry had been working at the Grosse Ile Baptist Church as the minister's secretary since sometime in 1981 and sometime in '82 they added her to the payroll. The job actually encompassed considerably more than the name implied. She worked forty hours per week from 9:00 A.M. to 5:00 P.M. and had to record the minutes of the Deacons and Elders meeting on Monday night, attend something else on Tuesday evening, Wednesday was the mid-week service, Thursday was the Sunday School Teacher's meeting, Friday and Saturday evenings usually had a social function at someone's home, and of course Sunday began with Sunday School followed by Sunday morning Church Services, then lunch at some restaurant in a group, and then Sunday evening Services.

Dave may have felt that he had been abandoned, but the schedule was no different from any of the prior two Baptist Churches that she had joined. Then she enrolled in evening classes in a Business School in the area.

One evening in '82 she awakened Dave as she rolled into bed around 11:30 P.M. and announced, "Dave, as soon as Terri reaches eighteen and graduates from High School I'm going to leave you."

Dave responded, "Aw, knock it off. I have to get up at 5:30 to shave, shower, eat breakfast, and be in the office by seven so that I can be in Physical Therapy by 3:30." He didn't believe her threat was very sincere. He didn't realize how stressful their lifestyle had become to her.

Some time in early '83 his friend and former student, Tony Malizia telephoned him at the office and suggested that they have lunch together the next day. Dave didn't need to be encouraged to agree to that and they met the next day. At the conclusion of the meal Tony said, "Dave, I'm a friend, so take this as a friend. Sherry called and was looking for an attorney to represent her in a divorce action. I told her that I'm a friend of both of you and I won't represent either one in such an action. I'll recommend marriage counseling and if that fails I'll recommend an attorney for each of you."

Dave thought that approach displayed integrity and a lot of common sense, so Sherry and he began ten weeks of marriage counseling. After three weeks Sherry gave it up. Dave completed it and still didn't seek a divorce attorney until returning from the annual "renewal of the relatives animosity" during the Christmas '83 holidays.

Upon return he started receiving the bills on the charge cards that inspired him to close all accounts and proceed with a very unpleasant and obviously destined action. The marriage was terminated in April 1985.

LII

IF IT'S NOT IMMORAL OR ILLEGAL DO IT AGAIN

His next flight in an airplane was on July 14, 1985 when he took a check ride with Ed Bennet, Chief Pilot for Grosse Ile Flight Service to convince himself that he was still capable of flying an airplane. His curiosity was pleasantly satisfied. He could still fly. He submitted to an FAA, 2nd Class Medical Examination by his longtime examiner, "Doc" Kovach and the report was forwarded to the FAA Medical Branch in Oklahoma City, Oklahoma and in October his Medical Certificate arrived in the mail. Dave thought he was on the way back.

Kim Kovach gave him 2.9 hours of dual instruction in his father's Decathlon and endorsed his Biennial Flight Review on October 26, 1985. He was then legally qualified to act as the Pilot in Command of single engine land airplanes. He was ready to resume flying, but his right leg was so weakened from the stress that it had been exposed to that he didn't believe he would ever again be a flight instructor.

Kim's endorsement in his logbook was the first by any of his former students and just for the memories Dave silently resolved to obtain an endorsement from each of his former students that were qualified to do so.

In their discussions about his next step in rebuilding himself as a pilot Kim recommended that Dave consider purchasing a "Clipped Wing Taylorcraft" airplane to use to rebuild his skills. Dave then telephoned

The Spirit's Journey

Forrest Barber, the Taylorcraft Company test pilot in Alliance, Ohio, discussed the issue with him and learned of the, Swick-T".

Anyone who sees a ‚Swick-T" will think, "Clipped Wing Taylorcraft". It is and it isn't. It is built from a Taylorcraft airframe, but the differences in the control system, engine installation, wing design, ailerons, and the rigging distinguishes the, "Swick-T" from the, "Clipped Wing Taylorcraft". Dave then spoke to Jim Swick on the telephone and decided that he would put an advertisement in, "Trade-a-Plane", an aviation trades publication which said, "I think I want a Swick-T, flyable. Dave McKenzie, 313-671-1837."

On Friday evening, November 30 the telephone rang and when Dave answered a voice replied, "What do you mean, you think you want a Swick-T?"

"Exactly what it says." Dave replied, "Do you have one?"

"Yes"

"Is it for sale?'

"No."

"Then why are you calling me?"

J.R. "Buzz" Hurt from Odessa, Texas replied, "Because I have one out back in the hangar that I haven't flown in about two years and I just wondered why I'm keeping it."

"How much would you want for it?"

"I don't know. How much would you be willing to pay?"

The next morning Dave called Jim Swick and discussed it. It was the first "Swick" that Jim had built and was the airplane used to certify the installation of the 100 H.P. Continental engine, then the 125 H.P., the 150 H.P., and then the 180 H.P. Lycoming engines. Finally in '82 the wings were showing the effects of hard aerobatics, so "Buzz" Hurt asked Jim Swick to replace the wings with the new metal spar design instead of the original wooden spars, and had the ailerons and aileron hinges redesigned. The aileron system was not certified, therefore the airplane was licensed in the, "Experimental – Research & Development" Category. Jim agreed to perform a pre-purchase inspection on the airplane and Dave was comfortable with everything, so far.

On December 2nd he called Mr. Hurt and made an offer that was refused. On the 4th Mr. Hurt called Dave with a counter request that Dave refused. On the 6th Dave called Mr. Hurt and presented a slightly higher offer than his first one, but not as high as Mr. Hurt requested. On the 7th Mr. Hurt called Dave and accepted the last offer.

Dave then advised "Buzz" Hurt of his discussion with Jim Swick and "Buzz" agreed to take the airplane to McKinney, Texas for Mr. Swick's inspection. Both Dave and "Buzz" agreed that Dave would accept the airplane in McKinney if it passed Jim Swick's inspection and relicensed it. When the owner of an experimental airplane is changed the airworthiness certificate is instantly invalid and must be reissued by the FAA.

Another Swick-T owner, Randy Henderson, a Captain for Southwest Airlines, flew "Swick-T" N-13-BZ from Odessa to McKinney, Texas. Jim Swick's son, Mike started an inspection and as the project progressed Dave periodically discussed it with him on the telephone. When it became obvious that the airplane's condition was going to meet requirements of the inspection Dave advised the Swick's that his days of hand propping an airplane were over and he wanted an alternator, starter, and battery installed. They agreed that it would be done.

The inspection and addition of the accessories was completed and on February 23, 1986 Dave "flew with Delta" to Dallas where Mike Swick met and drove him to Aero Country Airport at McKinney. Dave agreed with "Buzz" Hurt. It was a "ten foot airplane" meaning that if you're more than ten feet away from the airplane it looks great.

Closer than ten feet to it the observer will begin to see the scratches and bruises. It had a few.

The winds were blowing twenty-five knots per hour from the west and both Dave and the flight instructor who was going to give him a check-out in the airplane agreed that it would be advisable to wait until later in the afternoon for the winds to decrease and then perform the check flight. It allowed Dave about four hours to become acquainted with some of the "airport bums" who kept coming and going in and out of the Swick's hangar and he soon abandoned any objections to the wait.

By 4:00 P.M. the winds had decreased enough for Dave and Charlie Jirik to agree to fly the airplane. Dave had no difficulty with it and Mr. Jirik seemed to be quite competent as an instructor. After some airwork in the airplane and a few landings both of them felt that Dave would

The Spirit's Journey

be safe in the little "bird" and they returned to the Swick's hangar and everyone viewed the videotape of Dave's accident in the Great Lakes.

One of those observing was a retired Braniff Captain, Charlie Lamb who was accompanied with a reputation as a man to avoid a confrontation with. Dave saw no sign of that personality trait and it seemed that they liked each other.

Charlie Lamb was the first Braniff Captain to hold a 747 Type Rating on his Airline Transport Pilot Certificate and when the company purchased a 747 to begin a nonstop service from Dallas, Texas to Honolulu, Hawaii Captain Lamb was one of the three Captains rated in the airplane. The company's procedure was to fly the airplane nonstop to Honolulu and then because of FAA regulations pertaining to flight time and rest periods for airline crews the outbound crew from Dallas would RON (rest over night) while the crew that had been in Honolulu would fly it back to Dallas where the third crew would return to Honolulu in it. Three crews kept the airplane moving.

On a trip from Dallas Captain Lamb had become aware of a Head Stewardess (They're "Flight Attendants" now.) that had been rather belligerent and had adversely affected the morale of the rest of the Cabin Crew. She continued to criticize and berate the other members of the Cabin Crew after their arrival in Honolulu and their being assigned rooms in the hotel. Captain Lamb contacted the Stewardess and explained that she was the "Boss" of the Cabin Crew while they were in flight, but on the ground and in the hotel she should, "get off their case".

When they returned to Dallas the Stewardess contacted her "friend" the company Vice-President and complained about the behavior of Captain Lamb. The V-P contacted the company Chief Pilot who had to contact Captain Lamb and discuss it. Captain Lamb's response was to refuse to apologize to the Stewardess and when the message reached the V-P a letter was sent down through the "Chain of Command" demanding a letter from Captain Lamb to the Stewardess offering his apology. He refused.

The situation continued to degenerate until Captain Lamb declared that he was going on vacation to work on his cabin in Colorado. The company with just three Captains for their 747 was in a "tight spot".

It was finally resolved when the V-P wrote a letter to the Stewardess over Charlie Lamb's signature and asked him to endorse it. Captain Lamb refused and the V-P had to rewrite the letter more than once until Captain Lamb finally received one that he would endorse.

The story ended. The case was closed and Captain Charlie Lamb exited the situation unscathed. Dave never learned what became of the Stewardess, but he liked Captain Lamb.

Following the check ride in the Swick-T Dave was asked by Jim Swick to get in the airplane and Mike Swick took some measurements of Dave's foot locations relative to the heel brake pedals and built some new brake pedals. That required all of Monday and the first half-day of Tuesday. During that time Dave met Al Backstrom, the designer and builder of the famous flying wing glider and airplane that were built during the late 1960's. Gene Soucy's hangar was next door to the Swick's and of course Dave visited Nancy and Gene's hangar.

Dave remembered Nancy going around the several homes on Aero Country Airport rounding up the children for a trip to the Dallas Zoo. He asked her, "Why are you doing this? I thought this was a zoo here." Her quick and spontaneous reply was, "It is, but we want them to see other animals."

The only flight instruments in the Swick-T were the altimeter and the airspeed indicator. The only navigation instruments were the compass and the clock. It had an exhaust gas temperature gauge, a tachometer, an oil pressure gauge, and an oil temperature gauge for engine instruments. The newly installed electrical system had an ammeter and since the airplane was a serious aerobatic mount it had an accelerometer to measure and record the "g" forces it was exposed to. There were no electronic navigation aids and no radio for communications. It was a "simple airplane". No one on the airport knew how much fuel its tanks held or what the fuel burn per hour was or how fast the airplane cruised. As Dave spread some charts (in FAA language that means, "maps") on the hangar floor and drew a line across them more than one person watched him plan the first leg of the flight back to Grosse Ile, Michigan, a suburb on the south side of Detroit.

On the afternoon of February 25 he departed Aero Country at McKinney, Texas thinking that he would be comfortable and could live quite happily in a place and among people like he had found there. A year and a half later he went back.

He made his first stop enroute north at Muskogee, Oklahoma to refuel. He knew there was no need for refueling, but he wanted to obtain reliable information on the fuel consumption rate and obtain a clue to the approximate capacity of the fuel system in the airplane. He decided that his initial estimates were pretty close to the real numbers and he then flew to Carthage, Missouri where he borrowed the, "airport car" and went to a motel that had a "Continental Breakfast" to spend the night.

On the morning of the 26th he flew to Wentzville, Missouri thinking that he would not fly this airplane in winds any stronger than he was exposing himself to on this leg. The winds were out of the east at 30 knots and the runway at Wentzville was number 18 & 36. As he approached Wentzville he began to think that he would attempt the approach and then decide on final whether or not to actually land. If not he'd go to another airport in the St. Louis area with a runway that ran more nearly into the wind. He was pleasantly surprised when he discovered that the Swick-T was controllable in that much cross wind, landed and rolled out with no problems until he parked at the fuel pumps, got out of the airplane and attempted to walk. The wind kept blowing him over, so he decided to hangar the airplane right there until the winds subsided.

On the 28th he flew N-13-BZ to Kentland, Indiana and then to Grosse Ile, Michigan at an outside air temperature of 4 degrees Fahrenheit. Knowing that the standard adiabatic lapse rate is 3 ½ degrees per thousand feet of altitude he remained 700 to 800 feet AGL and learned a little about his tolerance of low temperatures on those two legs of the trip.

He lost the little feeling that he had recovered in his legs after breaking his back in 1980 and as he was passing Warsaw, Indiana he found that he couldn't write simple numbers on the chart when he passed check points on the trip. He knew what he wanted to write, but just couldn't get the fingers to make the required movements. When he was on final approach for runway 22 at Grosse Ile he thought, "Here's where I'll roll it up into a ball."

He touched down and just sat there holding the stick back and the throttle closed, letting the airplane roll out and dissipate the speed of its own volition. It stopped after a roll out approximately 900 feet long and he thought, "Wow! What an honest, nice, well behaved little airplane. I like this."

He taxied to the hangar, locked the airplane inside, and soon his son, Mike appeared driving the T'bird. They went home and Mike left to meet his friends at a local restaurant while Dave lay in the tub soaking in hot water. After an hour of the "warm soak" he got dressed, sat down with a cup of coffee, and answered the telephone when it rang.

Charlie Lamb asked, "How are you? How was the trip? We hadn't heard from you and were becoming concerned."

PART LIII

THE LOSS THAT ALL MUST ENDURE

During Jewel and Mac's last visit to Grosse Ile in the spring of '84 Jewel had entered the house, stopped, looked around the living room and said, "I'll start to clean this mess up tomorrow." She knew that the marriage of Sherry and Dave was in the process of being dissolved and understood that Dave did not expect to keep the house. The next morning she arose, had breakfast and then went to work until approximately 1:00 A.M. the next morning. She followed that schedule for the rest of the week.

One morning she and Dave were sharing a midnight snack at the dinette table and Dave asked, "How did you and dad meet?" She related the story as written in Part III and added, "He asked me to marry him on that first date. I should have known then he was a damned fool."

When the family had purchased the home on Grosse Ile, Dave and the two children enjoyed watching television in the finished playroom in the basement. He was on the couch in the playroom watching a movie on television on Saturday evening, March 23, 1986 when at approximately 10:00 P.M. the telephone ringing on the wall in the laundry room awoke him. He was pleased to hear his mother's voice on the instrument, but he braced himself to receive a message that no one looks forward to receiving.

Mac had suffered a stroke about two weeks before and was hospitalized in Pensacola, Florida. Jewel and Jimmy remained in Pensacola near the hospital for a week and then began to commute daily from Evergreen to be with him. Jewel reported that they had returned from Pensacola about two hours before and she had just finished the laundry. Mac had not improved nor had his condition worsened. She expected him to remain incapacitated and speechless for the rest of his life.

Of course Dave had been advised of Mac's condition within a couple of hours of his stroke and Dave knew that there was nothing that he could do to alleviate the situation. Even though he would have liked to have been there he knew that he would have only been a distraction and another load on Jewel, his mother, so he remained at home on Grosse Ile or in the office in Dearborn and made a check on Mac's condition every evening. Jewel sounded strong and in control of everything. She said, "Jimmy had gone out and would be back home shortly." That didn't surprise Dave and he had a strong feeling that there was still no need for him to go to Evergreen. Jewel ended the conversation wishing him a "Happy Birthday" and they agreed to make contact Sunday evening.

He hung up the telephone receiver and returned to his position on the couch without any regrets about the TV movie. Sometime during the 11:00 o'clock news report he fell asleep and was awakened around 1:30 A.M. by the telephone. It was his brother calling and said, "Dave, it's mother. I had to call the hospital a while ago and they sent an ambulance for her. I think she's gone."

Jimmy and Dave agreed that Jimmy would go to the hospital and they would talk again in a few hours. A few hours later Jimmy called and advised that she was on life support, but he didn't expect recovery. There was nothing for Dave to do, but wait. Later that afternoon "Skipper" Stacey, a life long friend of the family called and confirmed Jimmy's prognosis. Jewel was gone on March 24, 1986, Dave's forty-ninth birthday.

He flew to Montgomery where "Skipper" Stacey met him and drove to Evergreen where they arrived about 11:00 o'clock Monday morning. Dave had been dealing with the news fairly well until they drove into the backyard at 115 Magnolia Avenue and he looked at the neat landscaping in the clean yard and the neat manner that Jewel arranged everything in the carport. He entered the door to the lower level that was the area he had lived in from the age of 21 until he married at 26 and saw that it looked exactly the same. It was as though Jewel was still there and he knew that it couldn't have been any different because she had only been gone thirty-six hours at most.

He waited about five minutes and then went upstairs to face the family friends that were waiting for him in the kitchen. When he entered the kitchen everyone in the room suddenly fell silent and looked at him with no expression on their faces as though they were waiting for him to dictate a response to his arrival. The gravity of the situation struck as he attempted to look each of them in the eye and he had to look away in an attempt to conceal his grief even though he was sure that all of them knew what he was feeling. He knew that it was not a time when everyone would expect him to be capable of concealing his grief and he knew that any attempt to do so would only appear to be a pitiful effort to play the, "tough guy". He bit his lower lip, hung his head, hugged the ladies, and shook hands with the men. He went into the living room and sat down while the visitors waited for him to compose himself. Soon they began to drift into the room and discussions of what actions to take next and in what order began. Starting to make decisions, ask questions, and to talk about the future removed his thoughts from the sense of loss, grief, and loneliness, and replaced them with the inner strength to accept the loss and to begin planning a future. It was good to have the friends there that day.

On Tuesday, March 26 Mike flew to Atlanta from Detroit and met Terri who flew from Raleigh, South Carolina. The two of them then flew to Montgomery where "Skipper" Stacey made another trip from Evergreen and met them. Terri was attending a Junior College in Jackson, South Carolina and Mike was still living with his father.

March 27, 1986 Jewel was interred in Magnolia Cemetery at the south end of the street where she and Mac had rented their first apartment when they moved to Evergreen in 1939. During the service Dave looked at the run down house across the street where Addie Bea Smith and her second husband, J. D. Smith had lived and where Jimmy and he had spent the night when Mac's father died in 1951. On the west side of the cemetery he could see where a house once stood that was rented by Mamie and Lindsay Daniels. He and Jimmy had slept there more than once. After the house was destroyed the site had been used as the city garbage dump and when it was filled the dump was relocated to the northeast side of town and the City Dog Pound was constructed on the site. Approximately a half mile south of Magnolia Cemetery the City of Evergreen's sewerage treatment ponds are located, but they are not visible from the cemetery and no odors are noticeable. Jewel and Mac had selected the burial plot for themselves, Jimmy, Dave and the grandchildren, but Dave never felt that Magnolia Cemetery at the south end of Magnolia Avenue in Evergreen, Alabama was the site deserving to be the eternal resting place of Blaney Jewel Currence McKenzie.

PART LIV

ADJUSTING THE AIRPLANE TO COMPLY

He didn't fly the airplane again for two weeks when the outside air temperature had moderated slightly and three weeks after that he attempted his first aerobatics in the Swick-T. He wasn't comfortable with the way it entered spins and snap rolls were totally unacceptable. He knew when he accepted delivery of the airplane in McKinney, Texas that the weight and balance forms were incorrect and the FAA required placards were not displayed on the instrument panel and two inch high letters spelling, "EXPERIMENTAL were not on the doors, so he began some research and inspections of his own.

He found that the weight and balance forms were still appropriate for a Taylorcraft BC-12 with a 65 H.P. Continental engine installed. This one had a 180 H.P. Lycoming engine on the nose. In performing a weight and balance analysis on the airplane he found that the 6-pound weight in the rear of the fuselage should have been 16 pounds, so he changed it adding the 10 pounds. He also added all of the placards and symbols for the operating limitations and changed the registration number to N-88-TD, N for United States, 88 because it's the number of the Michigan Chapter of the International Aerobatic Club, and TD for "Terri Dawn", his daughter. He spent the next three weeks personally inspecting, upgrading, and making the airplane compliant with the FAA regulations and then with the records and logbooks in hand visited the FAA Maintenance and Engineering Inspection District Office at Willow-Run Airport.

The Inspector agreed with Dave and accepted everything except the category that the airplane was licensed in. He and Dave decided that it should have been licensed in the Experimental, Research & Development Category. When Dave left the FAA office he felt much more comfortable. He was also beginning to learn that in the seven different District Offices of the FAA there must be seven different ways to interpret the same regulations.

When he flew the airplane on April 13, 1986 the airplane seemed to perform aerobatics, especially the spins and snap rolls much better. The snap rolls were still not as satisfying to Dave as those he had performed with Alton Hesler on May 10, 1959 in a Luscombe at Laurel, Mississippi. He never knew why.

It is difficult to find a place to practice aerobatics in the Detroit area because of the density of the population near the city and the air space structure with its accompanying regulations and restrictions. Those requirements forced him to fly southwest of Grosse Ile for approximately forty-five miles to reach an area where he could legally practice aerobatics. Then he would practice for twenty minutes and then commute back to Grosse Ile Airport. He tried to practice at least three days per week after leaving the office and once per day on weekends. All of that depended on the weather being permissible of course.

Then on July 3, 1986, almost nine years after his last contest entry he entered the 1986 Michigan Regional Aerobatic Contest at Marlette, Michigan. He finished in 3rd place in the Sportsman Category Known flight sequence and misread the sequence card causing him to overlook the Immelman figure in the "Free" sequence. He finished the contest in 6th place and felt that he had almost recovered to being himself.

PART LV

A LESSON LEARNED

On August 8, 1986 he flew his first flight in the International Aerobatic Club World Championships at Fond-du-Lac, Wisconsin in very strong winds and allowed his concern for his position in the designated area called, "the Box" to cause him to get a score of "0" on the spin. That caused him to finish in 15th position out of 19 competitors on the "Known" sequence. The second flight was the "Free" sequence and while he was descending on the back side of the loop an Ercoupe airplane cruising northbound flew through the "Box" in front of him which caused some concern and degeneration in his performance of the figure. He finished in 8th position out of the 19 competitors. The most memorable things that he carried home from Fond-du-Lac that year was the newly established friendship with Barbara and Art Miller, both retired from American Airlines.

The next contest that he competed in was the, "Great Lakes Regional" at North Benton, Ohio beginning on August 29, 1986. Kim Kovach and Clarence Landoski were also going to compete in it, so the three of them agreed to fly to the site in formation and Kim Kovach offered to make the motel reservations for all three of them in one telephone call.

His son, Mike went with them for that one. What a memorable experience it degenerated into.

Of course the flight to North Benton was without incident except the parking ramp in front of the "T" hangars at Grosse Ile was in such poor condition that the right brake caliper housing was fractured on the Swick-T when taxiing out. The fracture was discovered during the technical inspection for entry into the contest. Fortunately, Forest Barber had another brake caliper assembly in his stock at Alliance and the delay caused by the repair only required about an hour. Dave then flew one practice flight and they gathered in the airport restaurant for dinner.

After dinner Kim excused himself from the table, went to the telephone booth and returned in a short time. He discovered that there was some kind of "Firearms Sportsman's" convention occurring the same weekend in the area and all of the motels were full. Ed Toland, one of their friends from their home, Chapter 88 gave them a ride in his van to search for lodging. It was finally found on the east side of Canton.

They were required to be at the contest briefing in the airport office at 7:00 A.M. the next morning and it was already 11:00 P.M. when they entered their rooms. Clarence and Kim shared one while Mike and Dave shared another. The room that Mike and Dave shared was not very clean. The towels were clean but ragged. The beds looked O.K. The carpet and curtains were dirty, but it was getting late and they would need to arise early in order to arrive at the airport in time for the briefing. In short, they had to take what they got.

It was 11:45 when they finally went to bed and then Dave could hear a very noisy party in progress. He called the office and complained to no avail. He was about to lose consciousness by 1:00 A.M. when an automobile with no muffler parked outside the door to their room with the engine left running. Proof that the driver and passenger exited the car was the sound of two doors slamming.

By 1:30 A.M. Dave got up, put on his pants and shoes, and headed for the door. He was very seriously considering shutting the car engine down and hiding the ignition keys when he heard the two doors on the vehicle slamming, the fenders rattling, the engine revving, and the sound of tires rolling in gravel accompanied by the whole cacophony of sound decreasing. He undressed and returned to bed.

It required at least another forty-five minutes for his pulse rate to slow and his blood pressure to decrease. He was considering radical acts of retribution to perform and upon whom to perform them when there came a very loud pounding of someone's fist on the door to the room accompanied by an extremely loud male voice bellowing, "Hurry up. Ain't you through, yet? I gotta' change the sheets."

At 5:00 A.M. he and Mike got up, shaved, showered, dressed, and joined Clarence and Kim outside waiting for Ed Toland to arrive and retrieve them. Dave probably had two hours of sleep that night.

They threw their baggage in the van and returned to the airport restaurant explaining to Ed why they were either going home or changing motels that night. The briefing and breakfast in the airport office building went well. When Dave flew the "Known" sequence he overlooked one figure on the sequence card that he used for reference that was attached to the instrument panel of the airplane. Not only did missing the figure earn him a score of "0", but the remaining three figures in the sequence were flown in the wrong direction which caused them to be scored, "0". The next flight was the "Free" and he finished in 2nd place in it. His cumulative score left him finishing the contest in "LAST" position.

Clarence in his Citabria, Kim in his dad's Decathlon, and Dave with Mike in his Swick-T were preparing to depart when Mary Lou Leverance made the comment, "This ought to be something to watch. A Polock, a Hungarian, and a Hillbilly flying formation attempting to lead each other home."

Competitive Aerobatics is a demanding motor sport that requires good physical conditioning, rest, and sobriety. It is terribly intolerant of disregarding the need for compliance with those prerequisites. It was a very good lesson to remember.

PART LVI

THE NATIONAL CHAMPIONSHIP

The logbook has entries dated September 21 through 22, 1986 indicating that he flew the Swick-T number N-88-SK from Grosse Ile, Michigan to Frankfort, Indiana to Festus, Missouri, then to Carthage, Missouri, then to McAllister, Oklahoma, and finally Denison, Texas in a total flying time of 8.8 hours to fly in the 1986 United States National Aerobatic Championships Contest. When he landed on runway 18 left at Grayson County Airport he was met by the "raggedest" appearing "follow me" cart that he had ever seen. There was Randy Henderson the Southwest Airlines Captain dressed "casually" driving it and waving his arm at Dave. Dave followed him to a hangar operated by Don Ort that looked like an indoor junkyard and inside were the airplanes owned and flown by the "big name" aerobatic stars in the United States. Pretty soon Dave McKenzie's Swick-T joined them in the hangar and the Technical Inspector soon appeared to inspect the airplane and found no problems with it. Within an hour Dave was accepted as a competitor in the USNATS. After all the years that he had read and heard about the USNATS he was going to be a part of the story. WOW!

Tuesday afternoon September 23 he flew in the Sportsman Known category and sequence finishing 15th out of 21 competitors. It was a disappointment, but if he flew well in the Free sequence which he usually did then the cumulative score might give him a finish in the top 10 positions. He wasn't quitting. Thursday afternoon September 25 his airplane was on the starting line in position for the Sportsman Free sequence to be flown as the next category and sequence. He was walking through the Washington Aero hangar on his way to the "powder room" in the rear of the hangar to prepare himself physically for the wait and to be ready to fly. Suddenly the cane in his left hand slipped on an oil spot on the floor that had leaked from the crankcase vent on one of the competitor's airplanes.

He awoke and recognized a young lady who was a pilot for Southwest Airlines who was holding a paper towel in the edge of the hairline above his right eye. It was very quiet in the hangar and there was about half a dozen people standing around him. Someone said, "The ambulance is on the way." The two fire trucks and the ambulance were kept in the fire station which was in the next building north of the Washington Aero hangar and someone had walked next door to advise them of the need for their services.

The attendant from the ambulance placed an inflatable splint on his right wrist and with the assistance of the driver and the medical attendant he entered the rear of the ambulance that transported him to the Grayson County Hospital where he was met by an orthopedic surgeon.

The examination revealed that he had fractured his right radius and ulna that the orthopedic surgeon reset by using a "Chinese finger splint" and then temporarily splinted with wooden strips. His last words were, "Go home and see your personal surgeon no later than Monday." He telephoned Washington Aero and another couple that had driven to the contest picked him up at the hospital and returned him to the airport where he found that other pilots had pushed his airplane back to Don Ort's hangar and put it away for him. He never learned who they were and consequently never thanked them for the favor.

At the conclusion of the day's events at the airport Frances and John Niergard from the Chicago area offered Dave a ride to the "Pool Party" with a detour by their motel room. John was a retired American Airlines Captain and Frances his wife was an ex-Olympic swimmer. John didn't compete in aerobatics, but was an experienced Certified National Aerobatic Judge and had been at Owosso, Michigan in 1980 when Dave suffered the accident in an air show at a neighboring airport. While on the way to their room Dave was asked what he knew about the "Pool Party" and if he thought Frances should wear a swimsuit.

Dave had only heard that the party was considered to be "memorable" and recommended that Frances wear a swimsuit under her dress. They drove to the Sherman Inn where Dave waited in the car while Frances and John prepared for the party.

The annual "Pool Party" for the contestants, officials, families, and friends was held at the Denison Holiday Inn beside the swimming pool and even though he was suffering some discomfort from the fractured wrist Dave enjoyed the hors de oeuvres and avoided consumption of any alcohol. He didn't know what reaction might occur because of the mixture with the sedatives he had been given through the intravenous injection in the ambulance and hospital a few hours before.

The food was eventually delivered and enjoyed as some friendships were made and others renewed. As the evening progressed the aura of camaraderie dominated the scene even though the participants had to remain physically prepared to meet the demands that would be made by participation in their sport the next day. Dave enjoyed the association with refined ladies and sophisticated gentlemen who totally enjoyed an evening's social event without exhibiting compromises in deportment supported by excessive association with a guest named, "Al K. Hol".

Just as had been anticipated by Dave a couple at the shallow end of the pool started wading and then Harley E----- from New Mexico began to tease Frances asking her if she could swim or would like to just wade while he swam. Dave couldn't hear most of the conversation, but soon Frances and Harley stood, walked to the edge of the pool and then hand in hand leaped into the deep end of it. John Niergard was smiling and a hint at a small laugh would occasionally escape his lips.

Frances removed both of her shoes and tossed them out of the pool. Harley removed his shoes and socks and tossed them out. He then removed his shirt and tossed it out. Even though their conversation could not be heard by the other party guests observation of Frances' and Harley's smiles led them to believe something was developing. Frances' skirt came into view from beneath the water and she threw it out of the pool. Harley's pants followed and then Frances' blouse was tossed out. The water was at Frances' neck and then Harley's shorts appeared above the water and flying towards the walkway. Both swimmers were smiling at each other as they turned ninety degrees and then dove beneath the ripples in the pool. At that moment and with that maneuver by the swimmers it became obvious that Frances was dressed in a full body swimsuit and Harley was in the pool and being viewed by approximately 150 party guests with a wardrobe consisting of his epidermis. John Niergard was laughing hysterically. Frances climbed out of the pool displaying a knowing smile. Harley kept swimming and would frequently rise and then surface dive revealing an excellent view of his glutinous maximus muscles each time he did so. After ten or fifteen minutes of his swimming exhibition someone threw his pants back in the pool.

During the ride back to Sherman, its Holiday Inn, and the Sherman Inn Frances thanked Dave for recommending that she wear a swimsuit beneath her street clothing. Dave suspected that John had probably seen such a display before.

PART LVII

A CASUAL ACQUAINTANCE

She was seated across the table from Dave at the Pool Party and he couldn't recall the exact details of their first meeting that had occurred the day before when he was attempting to find a ride to the Bar-B-Que dinner at the close of Thursday's activities. Bill Larson from Oklahoma had told him that the only thing he needed to do was stand by the driveway and "look pitiful". He'd be offered a ride. That's what Dave did and she offered a ride to the Bar-B-Que in her Cadillac. He assumed that she was going to attend the Bar-B-Que, but that was not her intention. She simply gave him a lift and he felt that he might have made some comment that she didn't approve of. Oh well, he thought that if he had it certainly was not intentional. So be it.

The next day he encountered her several times in the hangar and each time he tried to avoid remaining in her presence too long or too close for fear of boring her and then she shared the table with the Niergard's and he at the Pool Party. He suspected that she was simply displaying the socially acceptable degree of concern and sympathy for his injury.

A question arose about his plans for returning to Detroit and she said that she was planning to drive a friend to Dallas Saturday afternoon to catch a plane back to her home in Canada and she'd be happy to take him along. Dave thought she was simply trying to assist a person who had an obvious need and tried his best to accept the offer graciously.

The Spirit's Journey

Friday evening she offered him a ride to the awards banquet and he accepted. Of course they sat together during the banquet and he felt more at ease in her presence than he had before. At its conclusion he felt that she wanted to join the others in the lounge, but he explained that he was suffering some discomfort with his right wrist and felt that he should return to his room. He could detect some disappointment from her at that announcement and she offered to pick him up the next morning

Saturday morning, afternoon, and evening went as planned. She was good company and their relationship was as friends and acquaintances. They spent Saturday evening with some of her friends at an establishment called, "Judge Roy Bean's" to enjoy dinner, drinks, and conversation followed by Dave going home with one of her friends, a bachelor to spend the night.

Sunday afternoon she drove her Canadian friend and Dave to the airport west of Dallas and insisted on remaining with him in the coffee shop until his plane was loading for the departure to Detroit. The conversation was long and perhaps nothing really important was said or promised, but something was happening. There is an aura around people that cannot be seen, or heard. There are times when it simply grows and becomes impossible to not be noticed and difficult to ignore. In that airport coffee shop he felt that aura forming around them, surrounding, growing, and enveloping both of them. He was aware of it, feeling it, and enjoying it. She was probably unaware of suddenly becoming more than a "good Samaritan".

Yes, she was about three years older than he, but he was already nearly fifty and ladies were expected to live about six years longer than males, so discount that question. She smoked. He didn't. No one's absolutely perfect. She seemed to consume a few more cans of Coor's the day before than he, but she was obviously not an alcoholic. So what? He realized that he was deliberately judging her and wasn't finding any "negatives".

"Well, dummy," he thought, "you like this lady." There are times when some thoughts cannot be ignored and he thought, "Perhaps there's a small spark of humanity left, even in me.". They were at the gate when for some unknown reason he felt the strongest desire to hug her. Not as man/woman, but as very close friends. That description didn't really describe his feelings, either. He suddenly felt that he cared very much for this lady.

He leaned back against the railing so that they wouldn't fall and reached out to her right shoulder, gripped it firmly, but gently too so that she could shrug his hand away and tugged. With his right arm he hugged the lady that stepped toward him and felt the quick small buss on his left cheek. He immediately felt that he had failed to relay his feelings to her, so he did it again.

They parted and he could detect a few tears in her eyes.

Had he hurt her by being presumptuous? Did he insult her? Had he disappointed her? Was his attempt to hug her taken as an insult? The tears that he saw in her eyes told him that he wasn't alone in regretting this departure.

He boarded the plane to Detroit and his daughter Terri, her husband Bobby, and Mike were waiting for him at the gate with a wheel chair when he arrived.

Monday he went to the office and noticed that he seemed to be a distraction to the staff. Time reports were due to the supervisor Tri Moga and he presented Dave's report to Lee King, the leader and asked, "What's with McKenzie? Look at this mess. Did he forget how to write?"

Lee's response was, "Wait till you see him and then ask me again." Of course Mr. Moga could see and hear the humor in Lee's face and voice." He never challenged Dave about the quality of his signature. After all, Dave, who was right handed, was attempting to write left handed with his right wrist and arm in a splint and sling.

He did have to take a couple of hours off at noon to visit the Orthopedic Surgeon, be examined, and make the appointment for surgery on Friday morning.

That evening the telephone rang and he discovered when answering that IT WAS HER. IT WAS SHE.

PART LVIII

HOLD FAST TO DREAMS

His first contact with her had been in September 1979 when her husband had asked Dave to give her a check out in their son's Champion 7KCAB and some instruction in the tail-wheel airplane. Her husband was probably attempting to disarm any attempt by her to object to his hobby and its costs. In doing so he was carefully selecting the instructor seeking one who was not only proficient at flying but also competent at the profession of flight instructing. It is not a discredit to him that he was also probably looking for "as much as he could get for his money."

Dave met all of those qualifications. She and her son, Don, the owner of the Citabria arrived together on September 25, 1979 and Dave gave her one hour of instruction in the airplane that included some air work and concluded with three take offs and landings. The first hour of transition training from tri-cycle to tail-wheel airplanes is usually devoted to air work, but she could fly the airplane. Dave could see it, could feel it and was very pleasantly surprised.

They flew four times for a total of five and two tenths hours in the Citabria that included the introduction of spins to the right and to the left in the second hour. October 27 was the date that he endorsed her logbook as being competent to fly tail-wheel airplanes and capable of flying the Citabria solo. Although that isn't considered to be significant to the average non-pilot or to the "old time" pilot it is unique at this time in aviation history. He wasn't concerned about her safety. She "had it". The only thing she needed to do was continue to practice and he wondered if she had the one characteristic that would complete the psyche of a pilot, motivation. He thought that her substitute for motivation may have been pressure and encouragement from her husband and then she wanted training for an instrument rating.

Our country's military services and the airlines give pilot training candidates stanine tests to determine whether or not they have the psychological characteristics that will lead to completion of training. In the civilian arena we have no such luxury and a flight instructor must use his/her own intuition in deciding whether or not the investment of time and effort is justifiable in the particular candidate for the license or rating. Dave soon felt that she did indeed have the psychological characteristics that would enable her to become an instrument pilot and he was pleased that she was one of his students.

November 25, 1979 is recorded as the date that she began the flight training required to earn the instrument rating and they were using their Cessna Cardinal, model C-177-RG, registered as N-8261-G. During the training that extended over sixteen flights totaling twenty-eight and six tenths hours of dual instruction her husband was always in the back seat. Occasionally Dave would be commenting on the need for the use of more or less rudder pressure and would use the expression, "Remember Granny in the back."

She was amused by that comment and thought Dave was referring to her husband, but Dave was simply attempting to soften the criticism with the use of humor.

The only time that she annoyed him was when she was attempting to take the oral examination and flight test for the instrument rating with an FAA Inspector. Of course it's true that Dave would be happier and easier to associate with if he weren't so demanding and critical. He didn't like giving or receiving criticism, but it's part of a flight instructor's role in life and he didn't believe anyone had any hope for self improvement unless they subjected themselves to critique.

On the afternoon of the examinations she changed purses and overlooked transferring her medical certificate from the former to the current purse. The FAA Inspector had to cancel the exams, but he did not issue a violation of the FAR's because she had exercised the privileges of her Airman Certificate without having the accompanying medical certificate in her possession. She had to obtain her medical certificate prior to flying the airplane home.

She retrieved the medical certificate, rescheduled the exams with a "Designated Examiner" and earned the instrument rating the next day. She telephoned him at his office and advised him of her success and said, "When I landed I made what was probably the nicest landing that I've ever made."

He responded, "Well, we all get lucky sometime." As soon as he made the remark he was regretting it. Dave was annoyed with himself because he had not personally put the package of documents together for the tests. He had entrusted that duty to another person and he felt that it was a blot on his record that he'd always remember. Neither did he believe that pilots of airplanes should be dependant on "baby sitters".

They began work on a "Commercial Pilot Certificate" with two flights in May 1980, five in June, and one in July. The total was eight flights in eleven and one tenth hours of dual instruction in preparation for the oral exam and flight test, both of which she passed. Of course he was not surprised and was pleased. He didn't ride with her again for a long, long, time.

Memories of the period from July 13, 1980 through mid-November of that year were vague and he thought that it was in November that she and her husband visited him in the hospital room in Wyandotte General Hospital. There had been other visits, but he didn't recall them and didn't remember that she had been there. On that Saturday evening in Wyandotte she presented a wall hanging that she had made. She had airbrushed onto it the verse from Langston Hughes,

Hold Fast to Dreams,
For if Dreams Die,
Life is but a,
Broken Winged Bird,
That Cannot Fly.

At the time he thought she composed the verse that became such an inspiration to him.

A year later she and Kim Kovach arrived at his home for a visit and some time afterwards they had a very lengthy telephone conversation. Then in December 1982 they encountered each other again at Mettetal Airport in Plymouth, Michigan and at some time during the summer of '84 he saw her husband at the Ann Arbor Airport and learned that their marriage was in danger of foundering. On March 23, 1986 they met again at the Ann Arbor Airport and he was pleased to see that she and her husband were together and obviously trying to save their marriage. During that entire period he thought of her as "the other guy's wife". At no time had they even exchanged a handshake.

It must have been in April 1986 that several aviation enthusiasts were invited to a social function in a home in Pontiac, Michigan. He found himself sitting on a couch with a drink in one hand and not engaged in conversation with any of the other guests when he became aware of the arrival of another person. He couldn't identify the new arrival because the light behind her was only outlining the silhouette. As she strode deeper into the room and clear of the intense back lighting he recognized her, noted that she was alone, and surmised that her marriage was indeed troubled.

As she approached and spoke the sound of her voice confirmed that the figure approaching from the doorway was she. He stood and she continued forward until they met in an embrace. It was the first time that they had ever made a physical contact.

The embrace was not one of the passionate moves seen on television, in movies, or read of in books. It was the contact of two very good friends who hadn't seen each other in a great while and each was very happy to see the other. The surprise added to the pleasure. Suddenly he was very weak and almost staggering. Never had a simple contact had as much force and impact as that embrace. Oh yes, he had read the usual collection of novels and seen a lot of the movies, but they're not real. This was reality, not an act.

They separated from that embrace as friends do, joined the party, and were busy with everyone else for the duration of the evening.

Dave's experiences in Denison, Texas on September 25, 1986 led to far reaching effects that no one could possibly have anticipated. He had returned to the office on Sept. 29 and visited the orthopedic surgeon during the noon lunch break that day. The result was an appointment to present himself at 6:00 A.M. on Friday, October 3 for surgery on the right wrist. Wednesday evening his former student called and wanted to review on pylon eights in the Cardinal. He advised her of his incapacitation and agreed to fly with her but she would have to be the Pilot In Command on Thursday evening October 2 provided she could perform most of the labor to get the airplane out of the hangar at Ann Arbor. That was done and they shared dinner in Ann Arbor and he went home to spend the rest of the evening with his daughter, son, and son-in-law.

Friday morning the surgical repair was performed that included the installation of an external fixater to stabilize the radius and the ulna was pinned in place. Terri and Bobby provided the transportation to the hospital and then picked him up after the release later in the morning.

The next meetings of the former student and Dave were to attend the annual Christmas Party of the IAC in December and another to attend the Christmas Party of her part time employer at Willow-Run Airport. She had earned the Flight Instructor rating and it appeared to Dave that she was lonely and he believed that her marriage was irretrievably destined for failure.

On December 18 his new friend in Dallas met him at the Dallas-Fort Worth Airport and drove him to Denison to retrieve the Swick-T. Prior to his departure from the Grayson County Airport at Denison to return to Grosse Ile the two of them agreed to see each other again somewhere, sometime.

The husband that she had lost in an aerobatic practice accident was an ex-Navy pilot who had become a Captain with Braniff Airlines and one of the fringe benefits that his widow had was the privilege to fly on a standby status with any U.S. Carrier for the price of the tax on a passenger ticket. She could, would, and did make three trips to Detroit on visits with Dave and for a time he thought that the relationship was destined to have a future in defiance of the problems presented by the distance separating them. It wasn't possible.

PART LIX

1987, OPPORTUNITY DENIED

On January 13, 1987 he was scheduled to give Jerry Stanecki, a Special Features reporter for WXYZ-TV in Detroit an interview because of his comeback that was in progress after suffering the serious accident in 1980. He needed a pilot for the chase plane to carry Mr. Stanecki and the cameraperson for the demonstration during the interview, so he arranged for his former student to pilot the chase plane. There was too much snow on the ground to station Mr. Stanecki and the cameraperson at any readily available airports located beneath the appropriate airspace in the Detroit area to film a demonstration.

The interview went well and the flight for recording aerobatics air to air was also very smoothly executed and upon landing at Ann Arbor after the filming session Dave was met by Mr. Lee O'Berry, the Flight Standards District Office Manager and the Safety Programs Specialist for the FAA's Detroit Office. No violations of the FAR's had occurred and there was nothing carelessly or recklessly performed during the flight, so Dave had no reservations about his response when he heard of the FSDO Manager's allegations at a meeting the next night. He telephoned the FAA-FSDO and attempted to arrange a meeting to discuss the issue and Mr. O'Berry refused to have any meetings or to make any attempts to resolve the issue. Dave wrote letters to the Great Lakes Regional Office of the FAA, to Mr. Anthony Broderick, the FAA Administrator in Washington, D.C., his Congressman, and his Senator and never received any answers that took a stand on either side of the issue. That was the only time in his career as a pilot from September 1958 to date that any official has ever raised an issue about any of his operations.

He believes that every Federal Aviation Regulation is the result of an accident that resulted in a fatality. He recommends that pilots occasionally review the FAR's and remember that they are the consequence of indifference, carelessness, and ignorance. It is also his belief that the FAR's should be periodically reviewed by the issuing body because they sometimes grow unnecessarily lengthy, detailed, and voluminous which opens the door to confusion and rejection of them.

In August he entered the IAC World Championship Aerobatic Contest for the second time at Fond-du-Lac, Wisconsin finishing 15th in the Sportsman Known sequence and 20th in the Sportsman Free. It was a disappointment to him, but the airplane had not been flyable for most of the summer because Kim Kovach was recovering the empennage surfaces for him. Kim wasn't slow. The problem was that he didn't have access to the airplane that he should have enjoyed. The parking ramp that the "T" hangars were built on at Grosse Ile was being torn out and replaced. That would have left the airplane locked in the hangar unless he moved into Hangar 1 that was a huge bay hangar that had been constructed by the Navy during WW-II and he couldn't keep his tools, parts, and supplies with the airplane. It really limited his practice flights that summer.

The empennage needed recovering because during the time that "Buzz" Hurt owned the airplane the windshield had blown out of it while airborne, blew over the top of the fuselage, and damaged the top of the left horizontal stabilizer. The "mechanic" who was employed by Mr. Hurt had "repaired" it by coating an automobile body repair material called, "Bondo" on the fabric. Dave thought that mechanic was probably what a friend of his called, an "FAA Certified Money Eater".

His first contest for the '87 season had to be the International Aerobatic Club World Championship Contest at Fond-du-Lac, Wisconsin that he entered with only eight practice flights in his book. He finished in 19th place. He made eleven practice flights between August 15th and September 10th when he flew to Ottawa, Kansas to enter the Kansas Regional Aerobatic Contest and his new friend from Dallas exercised the privilege of her airline pass to fly into Topeka where he and a friend, Herb Hodge met her that evening in Herb's Cardinal RG and returned to Ottawa. He finished in 4th place in that contest and was not unhappy about those results.

A "Braniff Airlines" Captain that Dave enjoyed the privilege to become more acquainted with on that trip was Captain Ken Larson who was the first American pilot to hold the Concorde type rating on his ATP Certificate. In the late 1970's Braniff Airlines made an agreement with British Overseas Airways Corporation and Air France to have a Braniff crew meet the "Concorde" at its port of entry to the United States and fly the airplane to Dallas-Ft. Worth, unload it, and return to the east coast with another load of passengers. Dave considered Ken Larson to be "one interesting person" to be acquainted with. Another interesting person to appear at that contest was the 80-year-old retired TWA Captain Harold Neumann who was competing in the Sportsman Category flying his 1937 or '38 Monocoupe. Captain Neumann was one of the "big names" from air racing's "Golden Years", 1929 through 1939.

On Sunday afternoon, September 13 Dave and his "Texican" friend left Ottawa in the Swick-T loaded with both their sets of luggage and stopped in Muskogee, Oklahoma enroute to Addison Airport in Carrolton, Texas, a suburb on the north side of Dallas. It was a pleasant trip for Dave, but she had expressed some concern about sitting next to a door that was "all window". He forgot to cover it and she never made any comment about it. On the 14th he flew the airplane to Aero Country Airport at McKinney, Texas and hangared the airplane in Jim Swick's hangar.

Aero Country Airport was the home of IAC Chapter 24 and the entire week prior to the U.S. National Aerobatic Championship Contest is devoted to practice for the event. Dave was critiqued by Randy Henderson, probably the world's top Taylorcraft pilot since Duane Cole, and by Ken Larson who was a National Judge.

He was looking forward to participation in the 1987 "Nationals" at Grayson County Airport near Sherman, Texas. Even though he felt very good about his performance in the first flight he finished it in 15th position. One of the other pilots pointed out that none of the first ten pilots to fly had finished in the top ten positions. There seemed to have been some inexplicable change by the judges in their judging standards that may have relaxed slightly after flight number 10. He didn't record his standing in the Free Program, but his overall finish was in 15th, a disappointment to him. And so 1987's season of participation in Competition Aerobatic flying came to an end. It had been his hope for 1987 to be his last in the Sportsman Category, but the poorly timed, supervised, and inspected reconstruction project at Grosse Ile Airport had seriously compromised his ambitions.

PART LX

THE OTHER LOSS

At Ford Motor Company he was still attempting to develop his skills on the computer system to the same level of proficiency as drafting had been in expressing his ideas on product design features and he slowly became re-accustomed to life as a bachelor even though Mike still resided with him. He was lonely and probably socially handicapped by the shyness that he kept well hidden. He was determined to avoid making life-changing decisions by deliberately seeking a social life, but he was always confident that social activities would take care of themselves and he would find his niche.

Dave sometimes suspected that the stress that was endured by Jewel when Mac suffered the stroke in March of '86 and for the next three weeks was probably the cause of her massive coronary and he was really moved by the response from Mac when Jimmy and he told him of their loss of her while Mac was hospitalized in Pensacola.

Sometime early in February, '88 Dave received a telephone call from Jimmy advising that their dad had been admitted to the hospital in Evergreen and the prognosis was not an encouraging one. Approximately two weeks later Mac gave up the life that he loved and enjoyed so very much.

During the marriage of Jewel and Mac neither Dave nor Jimmy saw any exhibitions of the love that existed between them. Neither son knew why they had married nor why they had remained together for the fifty-two years prior to Jewel's death but somehow it was obvious to everyone acquainted with the two of them that theirs was a stable and solid union that lasted.

They had both taken the vow, "to love, honor, and cherish" seriously.

There was so much in Mac's life that Dave also loved and admired and had attempted to relive in his place, but there was no way that could be done. Mac's death was a tremendous loss for Dave, but he attempted to keep it hidden and thought that Mac would have expected his feelings to be kept hidden. Perhaps he did succeed in covering his feelings, but as he sat beneath the small canopy covering the grave site with the cold rain water leaking on him and observing the Masonic Burial Ceremony he couldn't avoid recalling the thoughts that ran through his mind when Jewel was interred beside the grave that his father's body was about to enter. Mac, like Jewel, deserved so much better than this site, but Jewel and Mac had selected and resigned themselves to it. Mac had succumbed to the call for his permanent rest on February 19, 1988, his son Jimmy's 47th birthday.

PART LXI

1988, HE HITS HIS STRIDE

On May 14 he made the first flight of N-88-TD in 1988 after completing its annual condition inspection and on May 22 he was one of the "Seminar Speakers" at the annual 99's Pinch Hitter course in Ann Arbor. As he was departing the airport the controller whose voice he recognized as belonging to the tower chief answered his call for taxi clearance addressing him as, "Cessna 88-TD". That was the same controller who had called the FSDO in January '87 and caused all of the "uproar" with the FAA FSDO Manager, Lee O'Berry.

When Dave completed the pre-takeoff checks, changed to the tower frequency, and advised that he was ready for takeoff the controller cleared, "Cessna 88 Tango Delta cleared for takeoff and straight out departure approved."

Dave couldn't resist advising, "88 Tango Delta is an EXPERIMENTAL TAYLORCRAFT. Was that clearance for me?"

The controller responded, "Affirm. What's experimental about it?"

Dave responded, "Short wings, big engine, unique controls, special ailerons, larger rudder, and modified landing gear." On the climb out after takeoff he held 70 MPH (Vy, best rate of climb speed) in the climb and had 1000 feet AGL (above ground level) when he crossed the departure end of the runway and the controller asked, "Was that takeoff necessary?"

Dave's response was, "Normal takeoff." Since he had started flying in '58 it was the third "grouchy" exchange that he ever had with Air Traffic

Control. Two of them were with Ann Arbor Tower. He was tempted to respond, "I'll fly the airplane, you fly the tower." but restrained himself and he was pleased that no comments were made or action was ever initiated towards him.

On the two prior occasions that he was less than pleased about communications with ATC the controllers had been student controllers and the "squabbles" had been insignificant. At "pilot's bull sessions" criticism of controller's comments and clearances often dominate the participants conversations. Such criticism usually bored Dave and he felt that it was usually the attempt of someone on an ego trip to impress the audience at the coffee pot.

On June 11, 1988 he flew in the Salem Regional Aerobatic Contest and in the Known Sequence finished in 5th out of 19 contestants. In the Free Sequence he finished in 10th and that put him in 9th overall. He wasn't pleased with his performance and knew that he needed expert critique from the ground. In competition aerobatics the pilot is judged by what the observer sees. Not how well he really flies the airplane. He started flying more under the critique of "Doc" and Kim Kovach and knew that it was going to show up in the scores that he earned later.

On July 1 through 3 he flew in the annual Michigan Regional contest at Marlette, Michigan and finished 5th in the Known and 2nd in the Free, which gave him, 4th place overall in the Sportsman Category. He left Marlette with his head up and eyes, "front and center".

On July 26, 1988 he took his former student, Ilene Hemingway, now a Certified Flight Instructor for a ride in the Swick-T to obtain her endorsement of his Biennial Flight Review. That was another endorsement that he was very pleased to have in his book and he was looking forward to the annual IAC World Aerobatic Championship Contest to be held in Fond-du-Lac, Wisconsin again during the second week of August.

Sunday, August 7 began when he was asked to fly in a formation of four airplanes for filming by ESPN-TV for broadcast nationally on their network. On August 9 he finished 4th in the first flight and 5th in the second flight. Both flights used the Known Sequence and then on the 12th he flew in the Sportsman UNKNOWN, the first time that an Unknown Sequence was flown by the Sportsman Category and completed the contest in SECOND PLACE, overall. (Thanks, "Doc" and Kim.)

When the final results were posted Jean Sorg, the editor of "Sport Aerobatics" magazine asked him what he thought about it and how he felt. His response was, "Not bad for a cripple flying an antique airplane." She published the remark in the article she wrote for "Sport Aerobatics" magazine.

On September 15 he flew from Grosse Ile, Michigan to Champaign, Illinois and then to Jefferson City, Missouri where his cousin met him and then they drove to her home in Fulton, Missouri. He hadn't seen his cousin since 1949 until she and her husband appeared at the wake prior to his father's funeral the preceding February. She gave him a very good tour of the city and he was especially fascinated with the church where Winston Churchill had delivered the speech in which the term, "Iron Curtain" was introduced.

The next day he proceeded to Ottawa, Kansas and entered the Kansas Regional Aerobatic Contest, which really provided some fond memories. Among them was the opportunity to become better acquainted with Harold Neumann, the famous pilot from the "Golden Years" of air racing, 1929 through 1939.

Harold Neumann at 81 years of age was competing in the Sportsman Category and it was the only time that Dave ever wished someone would finish the contest in a higher position than the one he earned. Harold Neumann was flying the Monocoupe monoplane that he had flown in air shows in the fifties and sixties. The original horizontally opposed engine had been replaced with a 145 horsepower seven cylinder Warner radial modified with the installation of a Bendix PS-5-C pressure carburetor to enable the engine to run in both upright and inverted positions. Unfortunately the engine did not have an accessory pad to mount an engine driven fuel pump and the PS-5 carburetor must have a fuel pressure of 9 P.S.I. in order to function. The engine wasn't equipped with a starter either.

Harold Neumann installed an electric fuel pump as the primary pump and would use a hand operated "wobble" pump to provide fuel pressure to start the engine and to serve as the emergency pump in the event of failure of the primary pump. His starting procedure was to "wobble up" the fuel pressure to 9 P.S.I., move the mixture control from "idle cut-off" to full rich and return it to "idle cut-off", turn the propeller through four blades by hand, "wobble up" another 9 P.S.I., move the mixture control to rich, return it to "idle cut-off", turn the propeller through four more blades, "wobble up" 9 P.S.I., move the mixture to rich, leave the mixture at full rich, and tie a string from the electric fuel pump switch routed up and over the top of some of the fuselage tubing to the door that was in an open position. He'd turn the magneto switches to the "on" position and with the throttle "cracked" (slightly open) he'd return to the nose of the airplane and swing the propeller. The engine would begin to run and the resultant blast of air flowing from the propeller would blow the door closed, pulling the string that would pull the lever on the fuel pump switch to the "on" position. He could then casually walk around the wing struts and climb into the airplane. It was encouraging to see that intelligence, knowledge of the machine, determination, and ingenuity can combine and overcome the limitations and restrictions introduced by an 81 year old body.

Mr. Neumann made an error in his first flight that lowered his score and at the end of the contest Dave finished in 2nd and Harold Neumann finished in 3rd. A "cripple flying an antique and an old man flying another antique" silenced the laughter that weekend.

On Sunday morning, the 18th the weather was marginal by the time Dave arrived at the airport and found that Patty Wagstaff had departed before the weather deteriorated. She had left a note on his Swick-T stating that they'd see him in Chickasha, Oklahoma if he joined them. Bill Larson from Oklahoma City had told him that if he wanted to join in the practice sessions in preparation for the USNATs he'd be, "made welcome". He had been considering flying down to Aero Country Airport at McKinney, Texas for the week.

He decided to wait for better weather that was forecast for Monday and even though an extra day in the motel was boring and psychologically uncomfortable he was glad he waited. The weather was really poor.

Bill Larson had been sitting close by Dave during the dinner on Saturday night and when Dave announced that he should proceed to seek a ride to the motel Bill had said, "Sit down. Don't worry about it. All you have to do is stand at the front door looking pitiful and somebody will feel sorry for you and take you." Bill was correct. On Monday he flew from Ottawa, Kansas to Stillwater, Oklahoma, refueled the airplane, raided the "pogey bait" and coffee machines, and then flew to Chickasha where Clint McHenry was taxiing out for a practice flight when he landed. They exchanged waves as they taxied past each other and Dave suddenly felt comfortable about the decision to go to Chickasha.

It didn't take long for him to learn that Chickasha was the base for the oil company that owned the Lockheed Vega, "Winnie May" that the one eyed pilot, Wiley Post made two flights around the world in and later performed some high altitude research in. It was during those research flights that Post unveiled the world's first "space suit". He also became better acquainted with two Unlimited Category pilots, Tom Jones and Ian Groom. Both of them were flying Pitts airplanes that year. The next year Tom Jones bought a Sukhoi 26, the Russian aerobatic monoplane in which he lost his life in an air show at Oklahoma City. In the spring of 2004 Ian Groom lost his life in a Sukhoi when he flat spun into the ocean near Miami, Florida while practicing for an air show.

He was fortunate to receive some critique from Patty Wagstaff, U.S. Women's Aerobatic Champion and was pleased when she commented, "I don't really see anything wrong with your flying and think you'll do very well at the Nationals." Dave found her friendly, a pleasant conversationalist, and of course an outstanding aerobatic pilot.

He left Chickasha on September 24th and flew to Grayson County Airport at Denison, Texas to enter the 1988 United States National Aerobatic Championships Contest where it was to be the third time he was to compete in the USNATs. He placed 7th in the first flight, 5th in the second flight, and finished in 6th place overall. He had hoped to do much better and had some questions about some decisions made by the Chief Judge, but 6th place is a long way from 1st and he didn't consider an argument to be worth the potential reward. He decided that it would be the last year that he would compete in the Sportsman Category anyhow.

He arrived at Grosse Ile on the afternoon of October 2 and considered the 1988 season to be in the history books and over. He had never been pleased with the way the Swick-T performed snap rolls. At least not in his hands, so he made a pair of "stall strips" and attached them to the leading edge of the wings and proceeded to practice the published Intermediate Category sequence for the 1989 season nine times before he retired the airplane for the '88 season on December 6th.

PART LXII

CONFIDENCE GROWS

The completion of the annual condition inspection and the first flight of the Swick-T in 1989 occurred on June 3rd. He flew three practice flights and on June 8 he entered the contest at Salem, Illinois knowing that he was starting unprepared for competition flying. He had not prepared a Free Sequence for the Intermediate Category that he had decided to compete in. That meant he had to design a flight sequence that is a diagram of the series of maneuvers he planned to present. The sequence had to meet the design requirements specified in the International Aerobatic Club rule book and be checked and approved by a National Judge. He had to submit his entry form, proof of membership in the sanctioning organization, proof of liability insurance coverage, eight copies of the Free Sequence judging form accompanied by eight copies of the diagram of maneuvers as viewed by the judges when flown from right to left, and eight copies showing the maneuvers when flown from left to right. He also needed to submit himself and the airplane to inspection by the Technical Committee. That meant that he was required to present his pilot's license, medical certificate, and logbook with a current Biennial Flight Review endorsement and entries proving currency of flight experience, the airplane registration and airworthiness, radio license, and a current packing and inspection record on the parachute. Upon receipt of the satisfactory documentation the Technical Committee physically inspected the airplane looking for any sub-standard or marginal mechanical conditions. When all of those requirements were complied with and he had paid the entry fee he was admitted to compete in the contest.

When designing the Free program he had difficulty keeping the number of maneuvers equal to or lower than the allowed number and still have the total "degree of difficulty" equal the total number allowable. In desperation he finally added a "roller" to the sequence. A "roller" is the slang name for a turn that is performed while rolling the airplane about its longitudinal axis. The "roller" is considered to be an Advanced Category maneuver and is not normally seen in the Intermediate Category.

When all of the paper work had been accepted, the inspections completed, and fees paid he was allowed to have his name placed on the practice list. He was not aware of it, but when his turn to perform a practice flight arrived there were a lot of pilots that stopped what they were doing to watch a "cripple perform three rolls in a two hundred seventy degree turn while flying a Taylorcraft."

In his first contest in 1989 that was also the first contest that he flew in the Intermediate Category he finished 6th in the Known program and 6th in the Free. In the Unknown program he finished 4th. The cumulative scores of his and the other competitors in the category placed him in 4th place overall at the conclusion of the contest. He felt much better when leaving the contest than he did when entering and recalled the words used by Henry Ford, II when he had encountered some criticism, "Don't explain. Don't complain."

Another Intermediate competitor in that contest that had been helpful to him without intention was a Canadian pilot, "Biff" Hamilton. "Biff" had been a fighter pilot in the RAF during World War II and was one of only three pilots that Dave knew that flew the airplane he considered to be the most beautiful airplane used in that conflict, the Spitfire.

After his retirement from the RAF "Biff" had moved to Canada and was a corporate pilot for the Shell Oil Company. Dave considered him to be the model of what the "English Gentleman" is thought to be.

June 16 through 18 he entered the Ohio Regional contest again, but this time in the Intermediate category and finished in 6th in the Known sequence, 8th in the Free, and 6th in the Unknown. The cumulative score left him 6th place overall for the category. He could think of nothing that justified the "poor" showing, but he really wasn't ashamed of his performance either.

Then on June 30 through July 2 at the annual Michigan Regional he finished 10th out of 11 competitors in the Known, 11th out of 12 in the Free because he rotated the wrong direction in the quarter roll on the down line in the hammerhead and finished the sequence going the wrong direction. In the Unknown he finished 7th and had a cumulative finish in 10th of the eleven competitors. He was really disappointed in the results from that one and began to wonder what on earth always adversely affected his performance in the Michigan contest.

In August he entered the IAC World Championship Contest at Fond-du-Lac, Wisconsin for the 4th consecutive year and finished in 12th, 15th, and 6th respectively in the three sequences resulting in finishing 12th overall. The performance may not have been considered a poor one, especially considering his handicap and the age of the airplane, but he still felt that he was not displaying his potential abilities.

On September 8, 9, and 10 he flew at the Mt. Pleasant, Michigan Regional finishing in 1st, 2nd, and 3rd places respectively which gave him a finish in 2nd place overall. Now, that was more like what he was pursuing.

In September he flew to Litchfield, Illinois to spend three days with Marilyn and Phil Sisson and to get critique and coaching from Phil. The result was to finish 14th, 17th, and 23rd in the three sequences that yielded 16th overall at the USNAT's in Sherman, Texas for 1989. He had made an error in the first figure of the Unknown sequence that taught him a big lesson.

The Judges took a break between the flights at the halfway point in the Unknown sequence schedule and he was the first pilot to fly after the break. During the break he relaxed and lost his concentration on the sequence. In the Known sequence the first figure was a full roll on a climbing line and in the Unknown that they were given at the Nationals the first figure was a half roll on a climbing line. He had practiced and presented the full roll on the climbing line for figure one all season and when the officials took their break immediately prior to his flight he had lost his concentration and performed a full roll instead of a half roll on the first figure. He then had to add a half roll to achieve the correct entry attitude for the second figure. That error gave him a zero on the first figure and an interruption penalty for adding a half roll to recover the correct attitude. He honestly felt that he had been one of the poorest performers on the field, but he was never known for having a lack of determination.

October 13 through 15 he participated in the contest at Robinson, Illinois and finished 2nd, 4th, and 2nd places respectively which allowed him to complete the contest with a finish in 2nd place. His former "coach", Phil Sisson witnessed those flights and he felt a slight degree of redemption and thought that the "Nationals" may have been just a "kick in the rear of the ego".

November 19 was the last day he flew the Swick-T in 1989 and he made the resolution to begin 1990 earlier in the year even if he had to suffer frostbite to do so.

PART LXIII

THE NEW OBJECTIVES IN SIGHT

The first flight in 1990 was on April 15th after completion of the annual condition inspection and he picked up Clarence Landoski's parachute and carried it along with his "chute" to Tecumseh for their inspection and repack cycles. During the return flight to Grosse Ile he performed his first aerobatic maneuvers for the year, a four-point hesitation roll, a half roll to inverted flight, and another half roll from inverted to upright flight. No problems appeared in the airplane's condition or performance. Sometimes he wondered what the justification was for disassembling an airplane that was functioning very well, inspecting its parts and assemblies, and reassembling it. Any mechanic is likely to overlook something when possibly interrupted by a telephone call or a visiting customer. Dave always used extra care, caution, and attention to details when being the first pilot of an airplane after an annual or a one hundred hour inspection.

He made another resolution on April 28th after giving a pilot friend of another pilot friend a ride in the Swick-T. Of course the friend of a friend wore a parachute as did Dave and it soon became obvious that it wasn't necessary. Dave considered the flight to be a familiarization flight for the other pilot and had planned to demonstrate some aerobatics in a very gentle manner, but while attempting to get the other pilot to perform some steep turns it became obvious that the man was in need of some retraining. During a steep turn to the left he became excited and broke the window out of the left door with his elbow.

After returning to the airport and landing the airplane Dave decided to cease giving people aerobatic rides. He had been doing it over the years in the hope that it would encourage others to become better and more complete pilots and hopefully, join in the aerobatic activities. It appeared that most of them just wanted a, "thrill ride". He thought, "Let 'em go to a carnival". The only time in his life that he ever had such a ride was with his father when he was eight years old and his mother had encouraged it.

He went through the "hassle" of filing a flight plan when he telephoned the FAA Flight Service Station for a weather briefing and to check the Notices to Airmen on May 18 prior to departing Grosse Ile on a flight to St. Catherines, Ontario, Canada to enter an aerobatic contest. He contacted FAA-FSS by radio in flight and opened his flight plan prior to crossing the International Border as required and enjoyed a very pleasant flight to St. Catherines.

The airport at St. Catherines, Ontario, Canada is a "towered airport" and as usual was not difficult to enter and land on. As he "rolled out" he advised the tower that he needed to go through customs and was directed to park at the base of the control tower to await the return of the Customs Officer. Pretty soon a Customs Officer walked up beside the airplane, performed his inspection that consisted of looking inside and asking, "What's in the bag?"

Dave said, "Two pairs of clean drawers and two T-shirts with a tooth brush, a razor, and shirt." The Customs Officer was in a good humor and asked if he was one of the "stunt pilots". Upon learning that he was entering the contest the Customs Officer pointed out the building that was going to be the headquarters for the contest and wished him good luck. Dave hadn't expected the border crossing in the "Experimental" airplane to be as easy and smooth as it turned out to be.

He remembered the story about the aerobatic pilot from Louisville, Kentucky called, "Cab Ugly" who was flying northward to enter the Michigan Contest in his Pitts Special biplane, got lost, saw an airport across the river on his right, landed, and was met by a "follow me" cart.

It was on a Friday morning prior to the 4th of July and the Canadians were celebrating their annual holiday, the "Freedom Festival" and were having an air show at the airport he landed on. Seeing the small red colored "Experimental" biplane, they thought he was one of the performers arriving and greeted him cordially, filled his fuel tank without charge, and told him to taxi to the area in front of the gathering crowd of spectators.

By that time he had discovered his where-a-bouts and realizing that he had crossed the border without the proper clearance, flight plan, etcetera and decided that he'd better keep his mouth shut, so he taxied out and continued on his way to the Michigan contest where he arrived without further incident and performed quite well. Nothing was ever heard from any "bureaurats" about it.

Dave scored 1st in the Known flight, 3rd in the Free, and 4th in the Unknown which placed him in 3rd place overall when the scores were added together. Sunday morning the weather reports were not favorable for a flight direct from St. Catherine's to Detroit, but he believed that he could fly southeastward to Buffalo, New York, land, go through Customs, and fly around the south side of Lake Erie to avoid the weather. There were two Pitts Specials that were going to have difficulty entering the USA, so they agreed to fly formation with him to Buffalo, land, and go through Customs.

The flight was smooth beneath a solid overcast with good visibility and light winds. It was the only time that Dave had seen Niagara Falls and they really didn't appear to be so impressive as he led the two Pitts pilots to Buffalo where he enjoyed calling the tower, "Good morning Buffalo. Experimental Taylorcraft, flight of three, eight miles west, one thousand five hundred with Bravo, landing."

They landed in a loose formation and received a directed taxi clearance to customs where the pleasure ended. They were told to remain in the airplanes until the Customs Official came out of the building. Thirty minutes passed, another airplane joined them, and then another. An hour passed and a pilot got out of his airplane and walked from plane to plane explaining that the Customs Official was not in his office, but would be sent out when he returned. Another half hour passed and the Official appeared. Of course he never offered an explanation for the delay and none of the pilots thought he should antagonize the Official by challenging him.

The Customs Office was on the second floor of the building where the FAA Flight Service Station was on the first floor. Dave thought he should obtain another weather briefing because of the time that had elapsed since the one he received in St. Catherine's.

The FSS Communicator told him, "The weather is marginal. VFR flight is not recommended." as he turned around and walked away. The display of arrogance was not what Dave wanted to see that morning, so: he said, "I'll decide whether or not I'll fly today. Not you. I've requested a briefing and you had better respond with one."

The Pitts pilots also received briefings and the three planes taxied out together, but one was going to Massachusetts, one to Delaware, and Dave was flying to the Detroit area, so they took off separately and each went his own way.

Dave landed at an airport on the east side of Cleveland named, "Lost Nation", purchased fuel, checked the weather again, and took off to fly to Grosse Ile. In the rain that he encountered he was down to 700' AGL and following U.S.-80 westward alternately slipping the airplane left and then right to see through the side windows because the windshield was covered with rain and telephone, radio, and television relay towers are not often found in freeway medians. He broke out of the weather somewhere around Sandusky, Ohio and enjoyed an easy trip around the western end of Lake Erie and northward to home. It wasn't a bad weekend. He had renewed his acquaintance with Biff Hamilton and an Unlimited Category pilot, Pat Crutchley.

Pat's home was in Great Britain and she supported her hobby by working in a hospital in Toronto as an Anesthesiologist. The Canadians call them, "Anesthetists". At one time she had worked in Atlanta, Georgia and was quite incensed about the term, "foreign alien" that was used in an article in the newspaper, "The Atlanta Constitution" referring to her. Whenever other pilots saw or heard her make an error they'd take advantage of the opportunity to ask, "Pat, what planet did you come from?"

On the Memorial Day weekend he flew to Covington, Tennessee landing at Indianapolis and Paducah, Kentucky where he had to wait for an improvement in the weather before continuing to participate in the "Rebel Regional" Aerobatic Contest where he finished in 2nd position. Again he thought, "Not bad for a crippled old man in an antique airplane." He visited his son, Mike in Guntersville, Alabama and his longtime friends Mildred and Grady Thrasher in Grant, Alabama before returning home.

Then the next contest was at the familiar Salem, Illinois airport on June 9 through 10 where he placed 1st in the Known flight 3rd in the Free flight, and 3rd in the Unknown flight. Buck Carrol from Memphis, a professional DC-10 pilot for the Federal Express Company finished in 1st with 3.2 points more than Dave. Neither of them had any complaints.

Forest Barber, the former test pilot for the "New" Taylorcraft Company in Alliance, Ohio was the gentleman who had told him to look for a Swick-T instead of a clipped winged Taylorcraft hosted the annual Taylorcraft Type Club Convention at Barber Field near Alliance during the second weekend in every July and Dave decided to attend in 1990. He knew that he would be asked to fly an aerobatic demonstration, so he decided to obtain another Letter of Low altitude Aerobatic Competency issued by the FAA for performances in public air shows. Principal Operations Inspector Vincent J. Scarpuzza met him at 9:00 A.M. on the morning of June 13 at a small grass field southwest of Ann Arbor, Michigan to observe a demonstration flight for the issuance of a, "Letter of Aerobatic Competency".

Mr. Scarpuzza's opening statement was, "Now, you used to have one of these letters and you hurt yourself using it. Did you learn anything?" He was accompanied with a reputation for being "grouchy", but that morning he didn't seem to be annoyed or upset at all. He was simply covering all of the thoughts, issues, and questions.

Dave suggested that since Mr. Scarpuzza had never seen him fly he would recommend that he be allowed to fly one demonstration above 1500' AGL and if it was satisfactory in Mr. Scarpuzza's opinion he would repeat the performance at the lower altitude. After completion of the flight Mr. Scarpuzza approved the reissuance of a, "Low Letter" and Dave was ready to attend the Taylorcraft Type Club Convention at Alliance.

The next contest was the, "Ohio Regional" sponsored by IAC Chapter 34 and was held at London, Ohio on the weekend of June 15 through 17 where he placed 4th in the Known flight, 2nd in the Free flight, and 2nd in the Unknown flight finishing in 2nd place overall. The year 1990 was shaping up to be a good one.

Dave McKenzie

After landing from the Free sequence he had sat down in a folding chair located in the hangar that the contest was using as a gathering area for the participants and an attractive lady was sitting in the chair beside the one he selected. She seemed to be especially attentive to the actions of a one armed man who was closely inspecting Dave's airplane. He soon learned that the one armed man was her husband, Brad Green, a United Airlines Captain who had lost his arm in a collision between his bicycle and a truck. She was a United Airlines Flight Attendant.

After inspecting the Swick-T Captain Green joined them with another chair and they had a very pleasant conversation during which Dave learned that Brad Green was flying multi-engine airplanes again, but under the Air Taxi rules, FAA Part 135 and was trying to figure out how Dave was performing aerobatics. After returning home Dave mailed Captain Green a copy of the biography of the Royal Air Force pilot, Douglas Bader who had lost both legs in a low altitude aerobatic performance in 1928 and had still become a Wing Commander and shot down twenty-one German airplanes in WW-II. The book was entitled, "Reach for the Sky".

Few people pity or discriminate against a, "cripple" if they see him continuing to perform his duties and to meet his obligations. They are more inclined to assist him. Both Dave and Brad Green had learned and shared that opinion.

Then the annual "Michigan Regional" sponsored by his home IAC Chapter 88 was held at Marlette, Michigan on June 29 through July 1 and the weather was so poor that only one flight was performed by each contestant, the Known flight. He inadvertently performed one figure behind a cloud that was between his airplane and the judges. They couldn't see the performance of the figure and the rules specified that in such an instance the figure would be given a score of "0". He finished the contest in 7th position out of 11 contestants. He wondered if aerobatic contests in Michigan were jinxed.

On July 7, 1990 he flew the demonstration at the annual Taylorcraft Type Club Convention and at the conclusion of his performance was given a copy of a videotape of the flight and was told that it had been the most impressive flight ever performed at the convention. He was pleased, but attempted to avoid exhibiting his pleasure. Of course Duane Cole who flew a clipped wing Taylorcraft in air shows from

1946 through the end of his career in the late '80's and was thought of as the "Dean of Air Show Pilots" and Randy Henderson from Plano, Texas who also flies a Swick-T have never appeared at the Taylorcraft Type Club Convention. Either of those gentlemen would really present a memorable performance. There are others who should be thought of in that same league. Perhaps air show pilots and the old western gun fighters share one characteristic. "It doesn't matter how good you are there is still someone out there somewhere who's better."

PART LXIV

JUNE 17, 1990

Howard Ebersole, Colonel, USAF Retired had flown B-25's and B-24's in the European Theater during World War II and F-51's and then the F-86 Sabre jet fighter in Korea during the Korean conflict. He was the only pilot Dave ever met who had shot down another jet in combat, a Chinese Air Force MIG-15. Colonel Ebersole asked him to attend the Sandhill Soaring Club meeting at the Ann Arbor Airport Terminal to invite them to participate in the annual Michigan Regional Aerobatic contest. The International Aerobatic Club had opened the aerobatic contests to gliders in 1990 thinking that there would be some interest and participation that might improve public acceptance and interest in the activity.

Dave descended the stairs into the meeting room in the basement of the building, chose a seat, sat down, and awaited the invitation to introduce himself and state his business to the club. In the third seat on his left sat a lady alone who returned his gaze, nodded her head, and then turned her attention to someone else who was attempting to speak to her.

During the welcoming of visitors to the meeting he was introduced and was given the opportunity to present his reasons for being in attendance at their meeting. At the conclusion of the meeting the lady turned to him and said, "We go up the street to Bennigan's Restaurant after these meetings and I hope you'll join us." They sat beside each other at the restaurant, but he was occupied by responding to the inquiries of the other glider club members about his activities.

They left the restaurant and each in their own car departed going eastward on I-94 and he blinked the lights of the car off and on once as he passed her.

On Saturday, June 21 he made an aerobatic practice flight in the Swick-T and after about thirty minutes of the aerobatics he was fatigued and decided to fly to Richmond field where the sailplanes were based and rest for awhile.

He chose to park the Swick-T beside a restored Taylorcraft that was parked beside a tree and bought a cup of coffee out of the dispenser that he found in the single room building that was obviously the office, sat down outside on the picnic table and was soon joined by the owner of the Taylorcraft. Ron and he introduced themselves and were discussing the subject of mutual interest, Taylorcraft airplanes when he heard a feminine voice saying, "Dave, I didn't know you were here. I was in front of the hangar and would have come over here sooner had I known it was you that landed."

Soon Ron and he decided to let her inspect both of the airplanes closely. While Ron was holding the door open on his and she was looking inside Dave thought, "She knows I fly aerobatics and she's interested in seeing some performed. I'm going to Marlette tomorrow for a club meeting and a practice session, so perhaps I should overcome my shyness and bashfulness and "round up my nerve and audacity" and when she and Ron returned to the picnic table he advised her of the plans for Sunday and offered her a ride with him in the Swick-T if she cared to go.

The weather was too poor for aerobatic practice on June 22, but IAC-88 was going to hold its meeting in Teal's Restaurant in Marlette anyway. Ceilings were low, the visibility was acceptable in spite of the very light rain, and the winds were strong, so he flew to Howell and when walking from the airplane to the airport office he noticed all of the flattened noses on the office windows. He walked in, closed the door, looked around, and asked, "Is anyone here who can sign off a cross-country?" (Student pilots must have their Instructor's endorsement in their logbook for any solo cross country flight they make and must obtain the endorsement of a witness at each landing they make on the flight.) That "broke the ice" with the "airport bums club" and he saw her sitting in one of the chairs along the west wall.

They flew to Marlette, rode with Kathy and Hugo Ritzenthaler to the restaurant, had lunch, attended the meeting, rode back to the airport with Kathy and Hugo, and flew through scattered light rain showers back to Howell.

They had a cup of coffee in the airport office building while engaging in casual conversation and Dave finally declared that he must return to Grosse Ile before the weather conditions worsened and again he "rounded up his courage, suppressed his usual shyness", and asked her out to dinner that coming Friday night.

He left the office after shaving and brushing his teeth, drove the forty-two miles to her home, and rang the doorbell. He could not have described her appearance to anyone, but he certainly recognized the girl who opened the door and invited him in. They drove to one of the better restaurants near Brighton, had a very pleasant dinner with neither of them feeling any need to hurry after which they returned to her home.

He had taken videotapes showing the air-show flight that he suffered the accident in, the one showing the interview with Jerry Stanecki of Channel 7 TV, and the videotape of the air-show flight at the Taylorcraft Type Club Fly-in. He thought that it would only be fair for her to know who he was and what he enjoyed before there was any chance for development of a relationship to begin. Of course they shared coffee and cake, an introduction to Abbey, her dog that seemed to accept him and more casual conversation.

When he was leaving she accompanied him to his car. He turned, leaned against the side of the car, gently reached toward her arms, softly pulled her towards him, raised his hands to each side of her face, and very gently and softly kissed her. He opened the door of the Thunderbird, got in, closed the door, started it, and backed out of her driveway. He knew that something significant was happening.

PART LXV

CONTINUED COMPETITION

On August 4 he departed from Grosse Ile in the Swick-T intending to enter the International Aerobatic Club Championships at Fond-du-Lac, Wisconsin for the fifth consecutive year and on August 6th he finished the 1990 Known sequence in 5th place out of the seventeen competitors vying for 1st place.

In the Free sequence he forgot to perform the second figure in the sequence and blamed it on being distracted by a spectator who was attempting to engage in conversation with him while he waited for his turn to fly the sequence. He knew better than to allow anyone to distract him prior to a competition, but he had allowed courtesy to a person who did not realize how important it is to avoid distracting a participant in a competitive event to cause him to make the error that placed him in 11th position out of the seventeen competitors. In the Unknown sequence he finished in 11th again and his final position was in 11th overall. He really felt the pain of disappointment with his performance in the contest and was unable to find any explanation for it. Of course, he recalled the error he had made in the Known sequence at the Nationals the prior September that was caused by the loss of concentration prior to the flight and the results of this contest just reaffirmed the conclusion that he had drawn in '89. Regardless of the opinions of the spectators, fellow competitors, and friends he decided that HE MUST DESIST in having any inter-action or dialogue with ANYONE prior to and during the sequence that he was competing in.

Competitive aerobatics will teach a pilot a lot about handling an airplane and also a few things about him/herself.

On the 8th and 9th of September he was scheduled to perform for the air shows at Marion, Ohio for the annual Mid-Eastern Regional Fly In and Nancy agreed to attend the Fly-In with him. On the morning of the 8th they flew from Grosse Ile to Marion and prior to the show that he was to perform the opening act in a couple that he had made friends with at the Ohio Contest in London, Ohio appeared, Mr. and Mrs. Brad Green. He really felt that it was a sincere compliment for them to drive from their home near Cleveland to Marion to see his performance and thought he had some new friends.

He was pleased with his performance on both days of the Fly-In and Nancy seemed to be comfortable with what she had observed.

Then on the week-end of September 15 through 16 they flew to Mt. Pleasant, Michigan where he entered the Regional Contest sponsored by Chapter 80, the Western Michigan Chapter of IAC where he finished in 1st place in the Known on the 15th and on Sunday, the 16th he finished the Free in 2nd place and the Unknown in 1st. That gave him a cumulative score for the entire contest that put him in 1st place overall. He was pleased and Nancy seemed to take it for granted.

For the last few years Nancy had been spending the last week of September in Myrtle Beach, South Carolina and asked Dave to go with her during the last week of September. For the prior four years he had competed in the U.S. National Aerobatic Championship Contest at Grayson County Airport near Sherman, Texas and was intending to repeat it again, so he refused to go to Myrtle Beach and invited Nancy to attend the "Nationals" with him. She accepted the invitation.

On Saturday, September 21 they left Grosse Ile on the flight to Sherman, Texas and had to stop and spend the night because of poor weather conditions. Dave thought the weather might force them to stop along the way after he received the telephone briefing from Flight Service prior to leaving. He had reached Rensselaer, Illinois when he decided that he had pushed the degenerating weather conditions as far as he dared to. They spent the night and the next day flew on to Hannibal, Missouri, Neosha, Missouri, and then to Sherman, Texas.

After landing at Grayson County Airport he taxied to Don Ort's hangar door where he and Nancy unloaded their luggage from the airplane, piled it onto the parking ramp, and then they pushed the airplane in the hangar.

Nancy could tell that they were among friends and was beginning to enjoy her new environment when a gentleman offered to give them a ride to the motel and asked, "Where's your baggage?"

She pointed to the pile and said, "Right there."

The gentleman asked, "What pieces?'

"All of it."

He responded, "I'll bring the car."

Nancy volunteered to work as the Boundary Judge for the Unlimited Category on Monday morning. Because it was her first attempt at participating in an aerobatic contest Jim Taylor accompanied her to the Boundary Judge's position and coached her through the first few flights. Jim Taylor had been the President of the International Aerobatic Club and earned his "honest living" as a First Officer on 747's flying from Chicago to Tokyo, Japan and return. He was very favorably impressed with Nancy's ability and integrity when one of the candidates for a position on the U.S. National Aerobatic Team flew part of a figure outside the "Box" and she "called him, out". Jim argued with her and she stood by her opinion that was correct. Nancy fit into the group at the "U.S. Nationals" quite well.

Dave's first flight in the contest occurred on Tuesday, the 24th and he was very disappointed when he finished 16th in the Known category. Obviously there was something peculiar about the judging, but it wasn't possible to find or prove anything was wrong. Aerobatic Judges are people, too. As such they are open to the same personal and subjective opinions as anyone else and so of the first ten competitors, none finished in the first ten places. They had to accept the scores.

On Thursday Dave finished in 7th position in the Free Program and the cumulative score moved him up to 10th in the overall standings at that point in the contest. The next day he flew the Unknown Program and finished in a very credible 2nd place. His final cumulative score for the 1990 U.S. National Aerobatic Championship Contest placed him in the SIXTH position.

As was written before, "It wasn't bad for a crippled pilot flying an antique airplane."

Saturday morning September 28 they departed Grayson County Airport, refueled in Springfield, Missouri, and Charleston, Illinois. Dave knew the weather ahead was marginal and was not surprised when he was forced to discontinue the journey at Kokomo, Indiana. The next afternoon they arrived at Grosse Ile and Dave felt just as he always did when returning from a week with friends on the aerobatic circuit. He didn't feel like he was returning home at the first sight of Grosse Ile and its airport. He felt like he was facing a radical change in his life. He sometimes wondered if he hadn't been intended to exist as a pilot, but was a victim of a fate that denied flight in his destiny. There was no way for him to know what Nancy was thinking. He didn't know whether she could appreciate the caliber of the unique people that she had met or not.

PART LXVI

AIR SHOW FOR THE SOARING CLUB

The next flight of Swick-T, N-88-TD was on October 6 when he flew it to Richmond Field, the base for the Sandhill Glider Club and the South Eastern Michigan Soaring Association at Gregory, Michigan. Nancy was a member of the glider clubs and Dave and IAC Chapter 88 had agreed to perform an aerobatic demonstration for a joint picnic and meeting with them. The weather was perfect except for very strong winds from the west. Their runway was 18-36 meaning it was headed north and south, so only Hugo Ritzenthaler and Dave flew that day. The gliders weren't even removed from the hangars.

There were two Inspectors assigned as Monitors for the performances assigned by the Detroit Flight Standards District Office of the Federal Aviation Administration and they brought their families along for the Saturday afternoon picnic. No problems occurred in the flights and everyone seemed to have enjoyed the afternoon, but the glider enthusiasts never did respond to the invitation by IAC-88 to join into the aerobatic activities and also gain more influence in the aviation community by establishing raphor with another group.

Dave returned N-88-TD to Grosse Ile Airport and proceeded to disassemble the airplane that "Buzz" Hurt had called, "a ten foot airplane." When Dave had purchased it from him in February '86 it looked great if you were more than ten feet from it. Any closer, it showed its bruises, dents, and scratches. Dave had flown the airplane pretty hard for six hundred thirty two and two tenths hours in the four and a half years that he had owned it and performed quite a bit of maintenance on it. He had originally purchased the airplane with the intention of using it to regain his skills as a pilot and intended to sell it when he had made more progress on the construction of the Hyperbipe airplane that he was building. He had enjoyed flying the Swick-T so much that he hadn't made any progress on the Hyperbipe, so he decided to retain ownership of the Swick-T, rebuild it, and have two airplanes. So his flying was halted as he began the next phase in his life.

PART LXVII

A NEW START IN LIFE

By December 1990 he had become accustomed to Nancy being in his life and felt very comfortable with her companionship. He was sure that he enjoyed, trusted, and perhaps was even in love with her. It was time for a test. He asked her to drive with him to "L.A." (Lower Alabama) during the Christmas and New Year's holidays and he was very pleased when she accepted the invitation. He intended to let her see where and what he had originated from and to introduce her to his son, and his brother.

He rented a trailer to haul the furniture from his son's former bedroom to Arab, Alabama for his home. While spending a couple of nights at Mike's home they visited the Air & Space Museum in Huntsville and his friend, Jan Hendrix. Then continued southward through the college town, Auburn where Dave had attended college and then onward to his hometown, Evergreen where he introduced her to his brother, Jimmy.

They continued the trip to Pensacola, Florida where they visited the Naval Air Museum and by then they had been traveling and visiting for four days.

They turned northward and drove to Monroeville, Alabama to visit one of Nancy's aunts who had moved there some years before, settled, became active in one of the local churches, divorced, and remarried. Dave had reason to visit Monroeville too. He has a friend living on the east side of Monroeville that he has known since 1958. Jennings F. Carter is a retired, "ag pilot" (crop duster/sprayer) and loves airplanes.

They had dinner with Nancy's Aunt Faye and the next morning they drove out to J. F. Carter's home for Dave to renew his acquaintance and so that Nancy could meet him. While sitting in the family room a gentleman arrived and was invited in by Carter simply raising his voice loud enough to be heard through the closed door. The door opened and the visitor entered, stopped, and started waving a finger at Nancy. Nancy stood, walked towards him, and said, "Uncle Dick."

Dave wondered, "What's going on here? What am I getting into?"

"Uncle Dick" was Nancy's Aunt Faye's former husband. Nancy and he remembered each other from the time that her Aunt and Uncle had lived in Michigan.

They drove back to Evergreen to visit Jimmy again and then traveled on to Montgomery where they spent the night before returning to Michigan.

One Friday evening in March after Nancy arrived at Dave's home on Grosse Ile he surprised her by suggesting that they give serious consideration to the possibility of becoming known as, "man and wife". They had known each other for seven months and he explained that he had not purchased an engagement ring because he felt very strongly that since the lady was going to wear the jewelry symbolic of the promise of a lifelong union she should have some input into the selection of it. He didn't tell her that if she refused to consider his proposal then he would have spent several thousand dollars needlessly because he certainly wasn't going to wear it. They went jewelry shopping for the next three weekends, but only to visit one jeweler on each trip.

In the meantime he had designed a spring aluminum landing gear for the Swick-T and the required modifications to the fuselage structure to mount it. He was hoping to improve the performance of the airplane by streamlining the landing gear and closing the holes in the bottom of the fuselage. He also cut the top off the vertical fin and then added a modification to the rudder making it aerodynamically and statically counter balanced while maintaining the profile and area of the original fin and rudder. Duane Cole had told him that when he flew the airplane he thought the rudder and aileron forces were out of balance. Dave wasn't aware of it because of the compromised muscle sensitivity and poor propreoperception in his legs as a result of his accident in 1980.

Jack Ellenbass in Holland, Michigan made the new landing gear legs and Kim Kovach performed the welding on the fuselage modifications for Dave and then Jack sand blasted and primed the fuselage structure.

While sand blasting the fuselage structure Jack signaled Dave to approach and inspect something. As he looked at the right lift strut fitting he could see the crack that had been revealed when the three coats of paint and one of primer had been stripped away by the sand blasting process. That fitting was GOING TO FAIL. When it failed the right wing would have folded upwards and Dave's exit from the failed airplane and subsequent successful use of his parachute is questionable. On every aerobatic flight that he had been performing he had been pulling eight g's (eight times the force of gravity) on the airplane.

Kim Kovach suggested that he look at a fiberglass engine cowling that Doug Dodge in Bay City, Michigan was marketing for the Pitts Special with the O-360 Lycoming engine.

The modified Taylorcraft named, "Swick-T" had an IO-360-A1A Lycoming engine rated at 180 H.P. on it instead of the original 65 H.P. Continental engine and the engine retains its size regardless of what airplane it's mounted on. Use of the new close fitting cowling required the design and construction of a new firewall made of .016" thick stainless steel that was six pounds lighter than the original galvanized steel firewall that was .026" thick. While all of that was happening the Hyperbipe project was laying dormant in the garage attached to Dave's home.

PARRT LXVIII

NANCY & DAVE

Nancy and Dave became Mr. and Mrs. David McKenzie at 5:00 o'clock on Saturday afternoon, May 30, 1992 in the Martha-Mary Chapel in Greenfield Village, Dearborn, Michigan within sight of the various offices that Dave had been stationed in during his employment at Ford Motor Company. The reception was held next-door to the Chapel in the Eagle Tavern with music performed by the, "New Old Stock" band. The band played nothing but string instruments WITHOUT electronic amplifiers and their first evening as Mr. and Mrs. McKenzie was spent in the Dearborn Inn, the first airport hotel in the United States and located across the street from the Ford Motor Company Dearborn Test Track that had originally been the Ford Airport, the first airport in the United States to have a paved runway, and the manufacturing site for the Ford Trimotor airplanes.

On Sunday, May 31 they began the honeymoon that included a tour of the Gettysburg Battlefield in Pennsylvania followed by a visit to Washington, D.C. where they saw the Library of Congress, the United States Capitol Building, the White House, the National Cemetery at Robert E. Lee's home, Arlington, the Smithsonian Museums, and the Smithsonian Air & Space Museum's restoration facility at Suitland, Maryland. Then they visited George Washington's home, Mount Vernon, and visited Jamestown, Virginia enroute to Williamsburg, Virginia where they spent two nights and one day. When departing Williamsburg they stopped to tour Thomas Jefferson's home, Monticello, kept the two dollar bill with President Jefferson's portrait on it that they were presented at the door and returned to Nancy's home in Brighton.

Nancy had informed Dave of the listing of some property that included a turf runway near Howell on the real estate market. They first inspected it on Valentine's Day in 1991 by flying over it. The following Saturday morning they inspected the property by actually entering it on foot. Then they met with the Township Supervisor and discussed their intentions asking what, if any objections should be anticipated. Dave had submitted a bid to purchase it, won the bid, and became the owner on April 21, 1991. He then sold his home on Grosse Ile during the second week of May, moved his furniture and other belongings into Nancy's home in Brighton and when they returned from the honeymoon they had a place to reside.

Construction began on September 8, 1992 of the new home with the attached hangar that Dave had designed during his lunch periods at Ford with the suggestions and approval from Nancy, of course. His first proposal had been for a simple rectangular building with a gable roof and a forty-foot wide door at the end facing the runway.

Nancy commented, "Well, that's the hangar." Then asked, "Where do we live?"

Dave responded, "We'll put a bed in one corner and a microwave oven in another corner. What else do you think we need?"

The final design was for a two-story house of the Grecian Colonial design with four columns supporting the front porch roof and the 2000 square foot hangar attached. The Grecian Colonial style house was what he decided he'd want to live in when he walked past the "Toliver Mansion" on Belleview Street on the way home from school when he was a child. They had no idea how much grief they were bringing upon themselves with the idea of designing and building a custom home in Michigan.

Because the square footage area of the home was to exceed 2600 square feet the State required that an architect make and stamp his approval on the drawings. Dave's response was to present his drawing to an architectural firm that made the final drawings and affixed the firm's "Stamp".

The State of Michigan through its Department of Commerce, Bureau of Licensing and Professional Regulation requires that contractors and homebuilders be licensed by the State before being allowed to construct buildings in the State of Michigan. Issuance of the Builder's License requires that an applicant pass a written examination on construction standards, Zoning and Building Restrictions, and Regulations applicable to the building trades and contracts. It soon became obvious to Nancy and Dave that the primary interest exhibited by the State was in the Licensing Fee.

They suffered a completely traumatic experience with the "Builder" that they hired to construct their home resulting in their having to hire an attorney to dismiss him, break the contract, and then begin litigation against him in a frustrated attempt to recover funds that he had fraudulently taken. They even had a lot of difficulty prosecuting him for forgery of supplier endorsements on several of the checks they had provided for payment to sub-contractors and suppliers and, of course they had to pay the suppliers and sub-contractors. The Livingston County Sheriff's Department would not pursue the case of forgery because the perpetrator resided in Wayne County. The Wayne County authorities wouldn't do anything because the crime victim resided in Livingston County. It was like watching the three monkeys at a table who all pointed towards each other when asked, "Who did it?" Dave finally had to contact the State's Attorney General's Office to inspire the State Police to finally do something.

In a hearing before an Administrative Law Judge in the Department of Commerce the "Builder's" license was revoked and he was ordered to refund the money that he had scammed and stolen from the McKenzies. In Wayne County Circuit Court he was also found guilty of forgery and was placed on probation while he was supposed to refund the money he had stolen by forging the checks. He repaid approximately $1500.00 of it and then disappeared and the Probation Department simply did not reply to Dave's telephone calls and letters of inquiry.

After dismissing the so-called, "builder" Nancy and Dave had to do their own building supervision and contracting. Nancy actually did most of the communicating with the contractors because Dave's office was fifty-two miles away in Dearborn. They finally moved into the new house the day before Thanksgiving 1993.

They wanted to share the pleasures from their new home and its accompanying runway in the side yard, so they hosted a "Corn Roast" for IAC-88 and EAA Chapter 384 during the summer of 1994 which became an annual event that EAA-113 joined in 1996. It was held annually for fourteen years. They also hosted IAC-88's annual Judge's School in their home and hangar for five consecutive years during that period.

McKenzie's Landing, Howell, Michigan (photo author's collection)

PART LXIX

CAREER'S END

Lee King had been promoted to replace Tri Moga who had retired in 1988 from Ford Motor Company. The department manager who had held a higher position in the company, but was apparently being moved down the chain of command was known for demanding that his supervisors carry a very heavy workload and he held an overbearing schedule of meetings that interfered with the work schedule that he requested progress reports on. Lee King was a very conscientious employee and the stress that was generated by the conflict in the manager's style and his style caused him to request placement on a temporary medical leave. In addition to the transition from manual drafting techniques to the use of computer graphics and the influx of younger hires with an emphasis on college degrees and no apparent concern for their professional interests, backgrounds, or knowledge of automobiles it became obvious to the older long-term employees that management had lost its focus on the product. Morale began to degenerate. More meetings were called and job assignments began to lose the appearance of being focused on the objective of continuing to be one of the major manufacturers of automobiles in the world.

The company had begun to investigate the possible transition from manual engineering design to a system named, "Computer Aided Design", (CAD) in the late 1970's. Ford had assigned five different CAD systems to its five North American Divisions hoping to allow the experience with the five systems to enable the management to select the most useful and by 1982 had selected "Product Development Graphics Systems" (PDGS) as the system to be used company wide. In the early 1990's many of the long-term employees retired instead of submitting to retraining for the company wide conversion to another system, Solids Dynamics Research Corp. (SDRC) as the company's primary system. It appeared that the engineering staff at Ford was constantly in training instead of designing automobiles.

The failing morale appeared to have begun between the times when Harold Poling, the company President retired in October 1987 and Alex Trotman, the Englishman who headed European Automotive Operations was named to replace him on January 1, 1993. There had been five years and two months that the company operated without a complete management team. The "grunts" couldn't comprehend the justification for promoting the head of a non-profitable Division, European Automotive Operations to the top position in the company management. The worst was yet to come.

After replacing two of the American products with the two European products named the Concourse and the Mystique that failed in the American market and following his retirement from Ford Motor Company on December 31, 1998 Alex Trotman joined the Board of Directors of Solids Dynamics Research Corporation and his replacement was another CEO from European Automotive Operations, Jacques Nasser. Mr. Nasser's errors in management and reckless squandering of company assets purchasing failing European companies became legion. The renowned Firestone-Bridgestone tire failure scandal in the late 90's and early 2000's also occurred on his "watch". Then in 2004 the company management decided to change CAD systems again. They chose KATIA, one of the four systems that had failed the selection process in 1980-81.

In 1994 the department that Dave was working in had been abolished. The reasoning given was that the company was going to concentrate its efforts on the design and production of trucks and their derivatives, Sport Utility Vehicles because they had a greater profit margin than was possible in the production of automobiles. Dave was given a Performance Review with a rating of, "Excellent" instead of the usual, "Excellent Plus". The reason given was that during the prior year he had two days absence due to illness.

Dave refused to concur with the Performance Review pointing out that the average annual absence rate for engineering employees was four and four tenths days and the average age of engineering employees was forty-two years old. He was fifty-eight years old and had two days absence in the prior year, therefore he would not endorse the review as being accurate, correct, and fair. He was then transferred to Truck Design and Engineering and given an Employee Planning rating of, "Tier 4" indicating the company had no plans for further development of the employee. Dave thought the message was as "subtle as a fart in a diver's helmet".

One month later he was telephoned by his former department manager and offered an "early retirement buy-out". Of course he asked where the phrase, "Early Retirement" came from when his record indicated that he was fifty-eight years old and had completed more than thirty years continuous service with the company.

Dave remained in "Truck" serving as a designer until mid 2000 when he was transferred to "Checking". On September 1, 2001 he was transferred back to "Car Engineering" as a checker, the job he had served in during his last six months at "Truck".

Then in October 2001 he was offered another, "buy out" that he refused and Mr. Nasser was encouraged to sever his relationship with Ford Motor Company and Bill Ford volunteered to fill the position for an annual salary of $1.00 assuming Mr. Nasser's company obligations on October 29, 2001.

Dave McKenzie had seen enough and when he was offered an "early retirement buy-out" in May 2005 he accepted a separation date of June 30, 2005. He had hoped to complete forty years of service, but the last eight years appeared to be filled with nothing more than frustration and disappointment on his part and indifference on the part of the management of the company. He had no enthusiasm for the products that he was aware of for the future either, so he left Ford Motor Company after having served thirty-nine years and eight months as an employee and one year and ten months as a temporary agency employee prior to that totaling forty one years and six months in Ford Motor Company, Product Design and Engineering.

Nancy's employer, Hometown Newspapers where she had served twenty-eight years and achieved the position of Division Controller

The Spirit's Journey

was purchased by the third largest media organization in the world, Gannett Communications in 2004. She remained with Gannett in the same position for two more years and retired in 2006.

It's true that they were not wealthy, but security was not questionable either. They were happy and accepted the challenges presented by the changing life styles without concern about them and were looking forward to the opportunities offered by the new freedom from restrictions and limitations associated with the obligations to an employer or the expectations and demands of any other parties.

His life had been productive and interesting. At least he believed it was and he could see the end of it approaching with the satisfaction of knowing that he was leaving no debts unpaid, no apologies left due, and almost all objectives met. Any objectives that he left incomplete would not be noticed by anyone else anyhow. The legacy that began in that infinitely small sphere containing all time, space, energy, dreams, emotions, ambitions, and Spirit had been carried far by Mac, Jewel, and then their sons through time and space and their descendants would do the same.

EPILOGUE – PART VII

John Patten graduated from Evergreen High School in the class of '58, joined the staff of Bell Air Service in Pensacola, Florida and attended The University of West Florida while earning his Commercial Pilot Certificate with Instrument and Multi-engine ratings on it, and then was employed by the State of Florida in their Department of Aviation as a pilot. Upon completion of his tour of duty as an air traffic controller in the U.S. Army he joined Delta Airlines and became Captain of Douglas DC-6, DC-7, DC-8, and DC-9 airplanes with FAA Type Ratings in each of course. He also has the type rating and flew as Captain in the McDonald Douglas MD-11, the Boeing 727 and finally, the Lockheed L-1011, his favorite airplane. He later became a check pilot on the L-1011 and an Instructor in the company's simulator training facility in Atlanta. He retired from Delta Airlines in 1997.

EPILOGUE – PART XXII

Grady Thrasher succumbed to natural causes at the age of 78 in 1994 and his loving, supporting wife, Mildred followed him as she always had in 2004.

EPILOGUE – PART XXV

Gil Baker, an engineer at General Motors, owned the Piper Tri-Pacer mentioned and built a wooden amphibian airplane powered with a Chevrolet "Corvair" engine that he had modified for the purpose. When Piper Aircraft Corporation started investigating possibly using a Corvair engine Gil Baker became the G. M. Liaison on the project.

Anna Main was really the person managing, "Big Beaver Airport". Her father had originally opened the airport and Anna and John, his children inherited it. Anna was inducted into the, Michigan Aviation Hall of Fame in later years after her death.

Ford Motor Company hired Rod Pharis in 1965 and in '68 or '69 he resigned and moved to San Jose, California. He and Sue divorced and he later became quite successful. His success is undoubtedly due to not only being intelligent, and competent, but he is also one of the most energetic and enthusiastic people a person will ever meet. He also started flying and earned a Commercial Pilot Certificate with Multi-Engine, and Instrument ratings on it.

EPILOGUE – PART XXVIII

Harold Krier, born April 4, 1922 had become one of the United State's premier air show and competition pilots during his career. In the process of developing his Great Lakes biplane into one of America's top air show mounts which attracted Dave's interest in the airplane he had acquired the original factory owned dies for producing the wing ribs for the airplane and modified them to stamp ribs from .032" aluminum instead of the original .024 thickness. Dave purchased the ribs for N-88-SK from Mr. Kreier.

He died on July 6, 1971 when he aparantly attempted to bail out of the Spinks, Model 10 aerobatic trainer during spin testing and his parachute failed to open completely prior to impact with the ground.

EPILOGUE—PART XXX

Bill Barber had been a flight instructor during WW-II, a Captain for Wisconsin Air, North Central Airlines, Republic Airlines, and Northwest Airlines. He was the Captain of the United States World Aerobatic Team in 1962 that represented the U.S. in the World Championship contest in Budapest, Hungary and had flown for the making of a couple of movies. He succumbed to prostate cancer on October 10, 1987 at home while accompanied by his family and friends.

The Pitts Special, "Little Stinker" is now the first airplane that a visitor will see as he/she is entering the Smithsonian Air & Space Museum's Steven F. Udvar-Hazy Center at Dulles Airport outside Washington, D.C..

The Tailwind, N-5747-N was advertised for sale a year after the flight to Florida and one of Steve Wittman's former students, Gene Zepp, Chief Pilot for General Motors Corp. bought it and flew it back to Detroit where it was inspected and a few minor repairs were made by the mechanics at GM's hangar on Detroit City Airport.

Shortly after take off on the way to Willow Run Airport for some radio work the propeller separated from the airplane and Mr. Zepp made the forced landing on the football field behind MacKenzie High School. Of course a football field is too small for operation of an airplane, but it was the only available open area. The airplane was nearly destroyed, but no one was injured.

That night Paul Poberezny, the founder of EAA called Mr. Zepp on the telephone and the next morning EAA sent a truck to Detroit to retrieve the pieces. It was rebuilt in the EAA Museum's aircraft restoration shops at Oshkosh and hangs on display today in the Airline Terminal Building at Wittman International Airport.

EPILOGUE – PART XXXI

In the seventy one years that the writer of this composition has been involved with aviation the ultimate destiny of the "Clipped Wing" Cub, N-26-SK is the only airplane known to him personally to have struck a house. It struck the house after a windstorm tore it loose from its tie downs, lifted it, and then dropped it through the roof of a neighboring house. Dave was reminded that the media is to be blamed for much of the public's panic and hysteria about such events.

EPILOGUE – PART XXXII

The "Woody Pusher" project was later sold to Dave Gustafson, a writer that contributed to "Sport Aviation" magazine. Gustafson wrote the first magazine article on the "Lakes" that was published in the magazine's September, 1977 issue. He later sold the "Woody Pusher" to someone else and it is not known whether it was ever completed or joined the list of 2,199 projects attempted out of every 2200 that never fly.

EPILOGUE – PART XXXIV

Following retirement Bonnie and Bill Unger purchased property on a small airport north of Grand Rapids, Michigan named, "Ojibwa" where they parked a house trailer, built a small "T" hangar, and kept a Luscombe airplane that Bill used for personal pleasure flying. Both are now deceased.

Their oldest son, Dan completed college, became a flight instructor in the U.S. Air Force, and later was a Captain for American Airlines. He is now retired and resides north of Detroit and occasionally flies a Pitts biplane that he built for use in aerobatic contests.

"The sign" has not survived. Bill owned another hangar at a small airport south of Detroit and one evening Danny was attempting a take off and as he crossed the intersection of the two runways the sign deflected downwards as the airplane's wheels rolled across the crowned surface of the crossing runway. The wheels and the sign became entangled, destroyed the sign, and ground looped the airplane into the adjacent cornfield. The airplane was undamaged.

Dan Unger is now living in retirement.

EPILOGUE – PART XLIII

Four of the eleven performers named in Part LXV are no longer with us. Their loss is deeply felt by all of the people who enjoy observing air shows and aerobatics enough to become familiar with their names, biographies, equipment, and performances.

At the same time all of the survivors study the fates of those that they've known in the sport and business, but not in the sensation or thrill-seeking manner that the non-participant might expect. Yes, they do indeed feel losses emotionally and sincerely sympathize with the families and friends that are left behind, but the survivors diligently pursue knowledge of the details of the events that claimed the lives of the pilots involved to avoid the piloting errors, mechanical failures, emotions that drive them, and the social pressures that come to bear upon them that can lead to an accident that all of them diligently attempt to avoid becoming casualties of.

Captain Bill Barber succumbed to prostate cancer at home in his bed on October 10, 1987 accompanied by his family and a few friends.

Duane Cole succumbed to natural causes in February 2004. He was considered to be the epitome of the "professional air show pilot".

Montaine Mallot and Daniel Heligoin, "The French Connection" died in a mid-air collision while practicing for the 2000 Air Show Season at their base in Florida on May 27, 2000.

Professor Bob Lyjak is retired from his position as a Mathematics Professor at the University of Michigan and as an air show pilot. His beloved Waco Taperwing is disassembled and stored at his home near Ann Arbor, Michigan.

The Spirit's Journey

Oscar Boesch is still alive and well residing in Canada.

Dave McKenzie resides in their home with his wife Nancy beside their privately owned runway near Howell, Michigan.

Danny Clisham and Jim Mynning reside near Ann Arbor, Michigan and are well. Both are still active in the air show business.

EPILOGUE – PART LXIV

Colonel Howard Ebersole married his companion, Jeanne and relocated to one of his prior stations, Alamagordo, New Mexico where he began his final rest in 2005.

EPILOGUE – PART LXV

The contender for a position on the U.S. World Aerobatic Team that Nancy called "out" of the box, Lee Manelski, a gentleman, an airline Captain, an aerobatic flight instructor, and winner of the Alternate position on the U.S. Aerobatic Team appreciated honest judging of his performances. Shortly after returning home from the Nationals Nancy and Dave were saddened to hear that he was the instructor that was flying the Pitts S-2 that was on its takeoff run when Mel Blank, Jr., the son of the man known in movies as the voice of "Woody Woodpecker", piloted a helicopter across the runway in front of him causing the deaths of Lee Manelski, his student, and Mel Blank, Jr. The passenger in the helicopter, Kirk Douglas, the actor survived with relatively minor injuries.

AFTERWORD

Nancy and Dave McKenzie now reside on twenty acres of land with a runway that is 2240 feet long beside their home near Howell, Michigan. The site is depicted on the Detroit Sectional, the Detroit Terminal, and The State of Michigan Aeronautical Charts (maps) and is named, "McKenzie's Landing" with the International Airport Identifier Code, "1mi5". The airplane hangar is attached to the house and contains the "Swick –T" airplane that Dave is rebuilding. It also houses the "Hyperbipe" airplane and a Model T Ford "Speedster" that he is building. Of course the required shop equipment and the accompanying hand tools are also there.

Nancy's son, Scott resides near Howell with his wife, a stepdaughter, and a stepson by his wife's former marriage. Dave's daughter, Terri resides in Montgomery, Alabama. Her daughter and youngest son reside with their fraternal grandparents in Evergreen, Alabama and her oldest son is serving in the United States Army. Dave's son, Michael and his wife, Rachel live in Birmingham, Alabama where he is employed in broadcast journalism.

"The Spirit's Journey" continues toward its destiny in the infinite future with them being used as the vehicles providing transportation, shelter, and the necessities of life.

FINIS